MAKING VALUE FOR AMERICA

Embracing the Future of Manufacturing, Technology, and Work

Committee on Foundational Best Practices for Making Value for America

Nicholas M. Donofrio and Kate S. Whitefoot, *Editors*

NATIONAL ACADEMY OF ENGINEERING
OF THE NATIONAL ACADEMIES

THE NATIONAL ACADEMIES PRESS
Washington, D.C.
www.nap.edu

THE NATIONAL ACADEMIES PRESS   500 Fifth Street, NW   Washington, DC 20001

NOTICE: This publication has been reviewed according to procedures approved by a National Academy of Engineering report review process. Publication of signed work signifies that it is judged a competent and useful contribution worthy of public consideration, but it does not imply endorsement of conclusion and recommendations by the National Academy of Engineering. The interpretations and conclusions in such publications are those of the authors and do not purport to represent the views of the council, officers, or staff of the National Academy of Engineering.

This study was supported by generous gifts from Robert A. Pritzker and the Robert Pritzker Family Foundation, and Gordon E. Moore, with additional support from Cummins, Boeing, IBM, Rockwell Collins, Xerox, Jon Rubinstein, Qualcomm, and Edward Horton. Any opinions, findings, conclusions, or recommendations expressed in this publication are those of the author(s) and do not necessarily reflect the views of the people or organizations that provided support for the project.

International Standard Book Number-13:   978-0-309-32653-7
International Standard Book Number-10:   0-309-32653-2
Library of Congress Control Number: 2015932679

Additional copies of this report are available for sale from the National Academies Press, 500 Fifth Street NW, Keck 360, Washington, DC 20001; (800) 624-6242 or (202) 334-3313; www.nap.edu/.

For more information about the National Academy of Engineering, visit the NAE home page at **www.nae.edu.**

THE NATIONAL ACADEMIES
Advisers to the Nation on Science, Engineering, and Medicine

The **National Academy of Sciences** is a private, nonprofit, self-perpetuating society of distinguished scholars engaged in scientific and engineering research, dedicated to the furtherance of science and technology and to their use for the general welfare. Upon the authority of the charter granted to it by the Congress in 1863, the Academy has a mandate that requires it to advise the federal government on scientific and technical matters. Dr. Ralph J. Cicerone is president of the National Academy of Sciences.

The **National Academy of Engineering** was established in 1964, under the charter of the National Academy of Sciences, as a parallel organization of outstanding engineers. It is autonomous in its administration and in the selection of its members, sharing with the National Academy of Sciences the responsibility for advising the federal government. The National Academy of Engineering also sponsors engineering programs aimed at meeting national needs, encourages education and research, and recognizes the superior achievements of engineers. Dr. C. D. Mote, Jr., is president of the National Academy of Engineering.

The **Institute of Medicine** was established in 1970 by the National Academy of Sciences to secure the services of eminent members of appropriate professions in the examination of policy matters pertaining to the health of the public. The Institute acts under the responsibility given to the National Academy of Sciences by its congressional charter to be an adviser to the federal government and, upon its own initiative, to identify issues of medical care, research, and education. Dr. Victor J. Dzau is president of the Institute of Medicine.

The **National Research Council** was organized by the National Academy of Sciences in 1916 to associate the broad community of science and technology with the Academy's purposes of furthering knowledge and advising the federal government. Functioning in accordance with general policies determined by the Academy, the Council has become the principal operating agency of both the National Academy of Sciences and the National Academy of Engineering in providing services to the government, the public, and the scientific and engineering communities. The Council is administered jointly by both Academies and the Institute of Medicine. Dr. Ralph J. Cicerone and Dr. C. D. Mote, Jr., are chair and vice chair, respectively, of the National Research Council.

www.national-academies.org

Far too much of our nation is waiting for new ways of working to arrive. We hear lots of rhetoric about how the nature of work will change, as if it relates to some unknown distant future. The fact is that it is happening now, and we need a broader recognition of this fact and policies and education that reflect it.

—Charles M. Vest

Acknowledgments

This report has been reviewed in draft form by individuals chosen for their diverse perspectives and technical expertise, in accordance with procedures approved by the National Academies' report review process. The purpose of this independent review is to provide candid and critical comments that will assist the institution in making its published report as sound as possible and to ensure that the report meets institutional standards for objectivity, evidence, and responsiveness to the study charge. The review comments and draft manuscript remain confidential to protect the integrity of the deliberative process. We thank the following individuals for their review of this report:

Norman R. Augustine, Retired, Lockheed Martin Corporation
Ann Bartel, Columbia Business School
J.D. Hoye, National Academy Foundation
Henry Kressel, Warburg Pincus LLC
Javier Miranda, US Census Bureau
Ed Morris, National Additive Manufacturing Innovation Institute
Andrew Reamer, George Washington University
Robert E. Schafrik, Retired, General Electric Aviation
Lydia Villa-Komaroff, Cytonome/ST, LLC

Although the reviewers listed above provided many constructive comments and suggestions, they were not asked to endorse the conclusions or recommendations nor did they see the final draft of the report before its release. The review of this report was overseen by Julia Phillips, Sandia National Laboratories. Appointed by the National Academy of Engineering, she was responsible for making certain that an independent examination of this report was carried out in accordance with institutional procedures and that all review comments were carefully considered. Responsibility for the final content of this report rests entirely with the authoring committee and the institution.

In addition to the study committee, NAE staff, and reviewers, many other individuals assisted in the development of this report. Gina Adam, UCSB PhD candidate, and Youngbok Ryu, Christine Mirzayan Science & Technology Policy Graduate Fellow, assisted in data collection and presentation. Clair Woolley, NAP Editorial Projects Manager, coordinated the production of the report, and Michele de la Menardiere designed the cover. We also thank the many experts who met with the committee to provide their insights on the numerous business, economic, educational, and policy issues central to the committee's charge.

Contents

xi

Executive Summary

Technological developments, reengineered operations, and economic forces are changing the way products and services are conceived, designed, made, distributed, and supported. Manufacturing or "making things" can no longer be considered separate from the *value chain*, the system of research and development, product design, software development and integration, and lifecycle service activities performed to deliver a valuable product or service to market. Businesses are focusing on this entire system to ensure that they are "making value" for their customers and are less likely to be disrupted by competitors or new technologies.

Furthermore, the convergence of these factors is causing major transformations that require the United States—and the companies that operate here—to carefully examine their abilities to innovate and capture the benefits. Businesses are experiencing increased competition from emerging economies around the world; while US-based businesses remain world leaders in many measures of innovation along the value chain, competitors from other countries are catching up quickly. The nature of work has also changed, causing jobs that consist of repetitive tasks to be disrupted by automation or offshoring to lower-cost providers. At the same time, developments in software and data collection are enabling businesses to better understand customers' needs, optimize design and production processes, and discover new market opportunities, which can generate increased demand and new enterprises that will create jobs.

It is important for businesses and communities across the United States to understand and respond to the changes affecting value chains for manufacturing and high-tech services not only because these activities account for a significant portion of the country's economic growth and middle-class jobs but also because the economy as a whole mirrors these changes. The same technologies that are transforming manufacturing and high-tech services are poised to transform enterprises in energy, health care, and education. And, by some estimates, 50 percent of jobs are ripe for disruption.

THE COMPLETE VALUE CHAIN

Business and policy leaders need a holistic understanding of the value chain in order to take effective action in response to the changing manufacturing and high-tech sectors. To systematically identify and successfully address customers' needs and capture higher returns, businesses must draw on in-house capabilities and external partners to carry out a set of interlinking activities spanning economic sectors. For example, in order to sell cars, automotive companies engage in research, product development, supply chain logistics, production, and sales, as well as pre- and postsale customer services such as maintenance, financing, and information and entertainment capabilities.

While companies have always been involved in a range of activities that cross economic sectors, it is increasingly difficult to recognize clear dividing lines between manufacturing, the production of software, and the provision of services in a company's product offerings. The service content provided by manufacturers has grown in importance, accounting for a larger proportion of total revenues in many industries. At the same time, companies primarily known for software and services have branched into providing manufactured products.

NEW OPPORTUNITIES FROM DIGITAL
TECHNOLOGIES AND DISTRIBUTED TOOLS

There is tremendous potential to improve productivity and create new demand and new businesses along the value chain through the infusion of software, data, and distributed tools. Developments in data collection and analytics, digital manufacturing, and crowd-sourcing have opened up a wealth of possibilities for companies and entrepreneurs to better understand customer needs and desires, optimize design and production processes, discover new market opportunities, and acquire new investment funds. Distributed tools such as cloud computing are lowering the barriers for potential entrepreneurs to start new businesses. And many businesses in diverse industries are creating new offerings by integrating systems of software, data, and manufactured products. In the pharmaceutical industry, for example, there is great opportunity to provide apps and services to help people understand when they should take their medicine and thus enhance their adherence to treatment. In the automotive industry, the expansion of software and information content incorporated in vehicles both enhances product performance and provides additional services to customers.

COMPETITION IN THE GLOBAL ECONOMY

Globalization and the development of emerging economies are increasing competition. While the United States remains a world leader along multiple indicators of research and production of high-tech manufactured goods and services, other countries are advancing rapidly. US-based companies face growing competition from emerging corporations around the world. Many US-based businesses have moved their operations abroad to take advantage of growing demand from emerging markets, easy access to capital, more efficient operations, established supply chains, particular workforce expertise, and tax advantages. Others are beginning to move some manufacturing operations back on-shore, as cost advantages sometimes erode and the loss of connection with other parts of the value chain, such as research and development and new product development, becomes more problematic.

Although the development of economies around the world has intensified competition, it also presents an enormous opportunity to expand demand for US goods and services, which may be increasingly important to drive economic growth. Indeed, for the past 30 years, the birth rate of new establishments in the United States across the value chain—in production, retail, and services—has been declining. Considering the significant role of new businesses in job creation and productivity growth, this is not a good sign. It underscores the importance for the United States to produce world-leading businesses to sustain its economy.

THE NATURE OF WORK

Over the past several decades, globalization and technological advances have changed not only the total demand for production workers but also the nature of production jobs. Manufacturing jobs that consist of handling and attaching parts by hand or other repetitive tasks are largely disappearing. Factory work in the United States is shifting to favor specialty skills in areas such as robotics-controlled maintenance, advanced composites, and radio-frequency identification of parts. At Boeing's plant in Everett, Washington, for example, workers control high-tech machines that use indoor GPS and laser-positioning systems to assemble the 787's advanced composite parts.

Advanced technologies and streamlined operations improve product quality and speed to market, thereby enhancing the competitiveness of US manufacturing operations in the global economy and providing higher-paying jobs. But because these advances make each worker more productive they also mean that fewer employees are needed to produce each car, airplane, or bottle of medications. Similar trends are occurring in other areas of the value chain and the broader economy, such as transportation, retail, education, and health care, and are likely to continue as advances in robotics and software enable machines to perform more complex tasks. The best bet to aid workers that have been left behind by these transformations is to advance their skills and create an effective ecosystem that continuously attracts and creates skilled jobs in all sectors of the economy.

IMPLICATIONS FOR ENTERPRISES AND COMMUNITIES IN THE UNITED STATES

To meet the challenges of a more competitive environment, and to take advantage of emerging opportunities, companies will need to adopt new approaches, which will include reengineering their operations and management systems in ways that improve productivity and speed to market.

While every business aims to optimize its operations for productivity, very few have implemented the advanced practices necessary to achieve leading productive operations. Employee training programs and collaborations across value chains and industries are needed to transfer the deep experience that can empower every worker to critique and improve operations. To ensure a sustainable stream of new products and services, companies will also need to leverage technology and talent to better understand customer needs and identify market opportunities. More individuals will need training in the skills and practices that will help them identify opportunities and execute the business models and resources needed to commercialize solutions. The ability to take advantage of burgeoning opportunities associated with the integration of systems of products, software, and data will require advanced computing and connectivity capabilities as well as a strong talent pipeline for software development, data analytics, and systems integration. Integrating systems across value chains presents significant opportunity for businesses focused on front-end activities as well as traditional manufacturers.

Policymakers, educators, and community leaders have important roles to play to ensure that the United States has an ecosystem that facilitates the adoption of best practices and attracts and creates businesses along the value chain and the broader economy. Just as American companies and communities reinvented themselves throughout the 19th and 20th centuries as the emergence of new technologies were coupled with the adoption of new business processes and investments in education and infrastructure, the current changes require forward-thinking leadership and action.

To prosper in the 21st century, US companies and communities must take action to upgrade America's ability to "make value." The committee has identified the following recommendations as a blueprint for these actions.

CREATING A PROSPEROUS PATH FORWARD: RECOMMENDED ACTIONS

Individual businesses can create value by coordinating value chains and optimizing operations.

- Businesses should establish training programs to prepare workers for modernized operations and invest in advancing the education of their low- and middle-skilled workforce.

- All businesses in value chains for manufacturing and lifecycle services should examine their business models and search for missed opportunities to leverage distributed tools and coordinate value chain systems to provide new products and services and improve productivity.
- Manufacturers should implement the principles and practices, such as Lean Production, necessary to enable employees to improve productivity and achieve continuous improvement.
- Researchers should further investigate and codify best practices for recognizing unmet needs and commercializing solutions, and effective methods of teaching them.

Collaborative actions are needed to improve the education and skills of the US workforce, particularly in manufacturing and high-tech value chains.

- Businesses, local school districts, labor, community colleges, and universities should form partnerships to help students graduate from high school, earn an associate's degree, and take part in continuing education in the workplace. State governments and Congress should provide tax incentives or other methods (e.g., formal mentoring, certification programs) to encourage investment and industry involvement in these education partnerships.
- To reduce financial barriers to the postsecondary education needed for jobs across the value chain, the cost-effectiveness of degrees at US universities and community colleges should be measured, publicized, and improved.
- Businesses, industry associations, and higher education institutions should work together to establish national skills certifications that are widely recognized by employers and count toward degree programs, and improve access for students and workers to gain these certifications.

Collaborative actions are needed to encourage the development of new businesses across manufacturing and high-tech value chains to stimulate innovation and job creation.

- The establishment of local innovation networks across the United States will foster the formation of new businesses and connect entrepreneurs to individuals, investors, tools, and institutions, both locally and around the world, that can help grow their businesses.

Certain fundamental areas need improvement on a federal basis to facilitate innovation throughout the value chain.

- US programs that contribute to innovation, such as the Small Business Administration, Manufacturing Extension Partnership, and the National Network of Manufacturing Innovation, should be directed and optimized as appropriate to facilitate the adoption of best practices and help young businesses to grow.
- US infrastructure must be upgraded. Businesses across manufacturing and high-tech value chains must have access to reliable energy and natural resources, transportation, and communication systems. Increasingly, many businesses also need access to high-performance computing grids and information storage. A world-leading infrastructure will attract businesses and facilitate the creation of new ones in the United States.
- US fiscal policy must incentivize long-term capital investments. The current tax structure encourages a preference for quicker returns over long-term investments to create new products and businesses.
- Data suggest that the rate of new business creation in the United States is declining. To understand the causes of this decline and enable the formulation of policies to reverse it, the National Science Foundation and other research funders should put a priority on supporting studies in this area.
- Federal programs and statistics should be modernized to account for the diminishing distinction and complex relationships between manufacturing, information, and services.

In today's highly globalized economy, companies need the best teams in the world to stay competitive. This requires not only developing and attracting top talent but also leveraging diversity to achieve better team performance.

- Businesses and universities should implement programs to attract and retain diverse talent, including along gender, race, and socioeconomic lines. Diverse teams have been shown to be more innovative and often produce better outcomes.
- Middle schools, high schools, universities, and local communities should expand opportunities for students to participate in team-based design experiences and learn to use emerging tools that enable new business creation. Students exposed to these types of experiences are better prepared to innovate in today's economy.
- Congress must reform immigration policy to welcome and retain high-skilled individuals with advanced science, technology, engineering, and mathematics (STEM) degrees, especially those educated in the United States. Many of these individuals become entrepreneurs and the United States should ensure that their businesses are in this country.

TABLE ES-1 Recommendations organized by actor

Actor	Recommendations
Businesses	• Companies should examine their business models to search for missed opportunities to leverage distributed tools and coordinate manufacturing and product lifecycle services. • Businesses should establish training programs to prepare workers for modernized operations and invest in advancing the education of their low- and middle-skilled workforce. • Manufacturers should implement principles and practices such as Lean Manufacturing that enable employees to improve productivity and achieve continuous improvement. • Businesses should work with local school districts, community colleges, and universities to form partnerships to help students graduate from high school, earn an associate's or bachelor's degree, and take part in continuing education in the workplace. • Businesses should work with industry associations and higher education institutions to (1) establish national skills certifications that are widely recognized by employers and count toward degree programs, and (2) improve access for students and workers to gain these certifications. • Businesses should attract and implement programs to retain diverse workers, including along gender, race, and socioeconomic background.
Federal government	• Federal agencies and interagency offices such as the Advanced Manufacturing National Program Office should convene stakeholders to identify and spread best practices for value creation. • Congress should establish incentives for businesses to invest and be involved in education programs. • Congress must reform immigration policy to welcome and retain high-skilled individuals with advanced STEM degrees, especially those educated in the United States. • The Small Business Administration should focus on helping young businesses become globally competitive as opposed to focusing on older, established small businesses. • The National Science Foundation and other research funders should put a priority on research to understand the declining rate of new business creation. • Federal programs that contribute to innovation should be directed and optimized as appropriate to assist software and service providers as well as manufacturers. • Congress should modify the capital gains tax rates to incentivize holding stocks for five years, ten years, and longer.

continued

TABLE ES-1 Continued

Actor	Recommendations
Federal government (continued)	• Congress should make the research-and-development tax credit permanent to encourage businesses to adopt longer-term horizons in their investment decisions. • Federal agencies should facilitate industry and government cooperation to identify shared opportunities to invest in precompetitive research in long-term, capital-intensive fields. • Congress should support state legislatures and local governments to invest in a world-leading wireless infrastructure. • Federal information technology and computing programs should facilitate access to a world-leading infrastructure for high-performance computing. • Federal agencies should develop methods of accounting for the complex relationships between manufacturing, services, and information and consider multiple ways of collecting and organizing national statistics.
State governments	• State governments should establish incentives for businesses to invest and be involved in education programs. • State governments should partner with local governments, industry, higher education, investors, and economic development organizations to create local innovation networks. • State governments should work with local governments to optimize the decision-making process for urban development investments and siting to facilitate the creation of innovation networks. • State legislatures, with local government and Congressional support, should invest in a world-leading wireless infrastructure.

TABLE ES-1 Continued

Actor	Recommendations
Local governments	• Local school districts should work with businesses and community colleges to form partnerships to help students graduate from high school, earn an associate's or bachelor's degree, and take part in continuing education in the workplace. • Metro area governments should partner with state governments, industry, higher education, investors, and economic development organizations to create local innovation networks. • Metro area governments should work with state governments to optimize the decision-making process for urban development investments and siting in order to facilitate the creation of innovation networks. • Local governments, with state government and Congressional support, should invest in a world-leading wireless infrastructure.
Education institutions	• Community colleges and universities should partner with local school districts and businesses to help students graduate from high school, earn an associate's or bachelor's degree, and take part in continuing education in the workplace. • Middle schools, high schools, and local communities should provide opportunities for students to participate in team-based engineering design experiences and learn how to use emerging tools that enable new business creation. • Universities and community colleges should improve the cost-effectiveness of higher education. • Higher education institutions should work together with businesses and industry associations to (1) establish national skills certifications that are widely recognized by employers and count toward degree programs, and (2) improve access for students and workers to gain these certifications. • Universities and community colleges should act to improve the inclusion of traditionally underrepresented groups in science, technology, engineering, and mathematics (STEM) fields as well as other disciplines required for value creation, such as market analysis and design.
Other actors	• Researchers should further investigate and codify best practices for innovation and develop effective methods of teaching them. • University rating organizations should track and make transparent the cost-effectiveness of degrees at higher education institutions.

Prologue

Twenty years ago the Eastman Kodak Company and Fujifilm were facing very similar situations. Both companies received the majority of their income from the sale of photographic film, and both saw a revolution coming that would destroy the market for that film. With the development of digital cameras there would be less and less use for film. Both companies recognized this, but they responded in very different ways.

Kodak's efforts through the 1990s and the following decade have been analyzed extensively, and various theories and explanations have been offered for its choices, but the bottom line is that the company failed to find anything that could replace film. Through the 1970s and 1980s, as it increasingly focused on its most profitable product—film—it either exited or failed to enter a number of other areas that could have helped it adapt to the coming crash of the film market. As the rise of digital photography proceeded, Kodak experimented with various products to augment its film business, including hybrid digital-film cameras and pharmaceutical drugs, but nothing developed into a major market. After filing for Chapter 11 bankruptcy in 2012, Kodak shed many of its liabilities along with many of its product lines—including consumer cameras and film—and emerged in September 2013 as a much smaller and different firm. The company now focuses on printing technologies used by businesses and sensors used on touch screens.

The effects of Kodak's choices reverberated far beyond the company. Its home city of Rochester, New York, was hit hard economically as the number of people employed by Kodak in Rochester dropped from 62,000 in the 1980s to fewer than 7,000 in 2012. This played a large role in the dramatic decline in Rochester's population—from a peak of 330,000 in 1950 to around 210,000 in 2012—and in the drop in the average income in Rochester from above the national average to below.

By contrast, Fujifilm moved much more decisively into new product lines. Recognizing that it had extensive expertise in dealing with the antioxidant chemicals used in photography, it used that expertise to develop antioxidants for use in cosmetics that help improve skin condition and now has a major cosmetics line. It developed optical films for use with flat-panel screens. Indeed, it became so accomplished at developing new technologies that in both 2012 and 2013 it was named one of the top 100 most innovative companies in the world by Thomson Reuters. It remains a strong and profitable company even though its previous major product, photographic film, now accounts for only a tiny percentage of its sales.

At its core, the tale of Kodak and Fujifilm is a story about the importance of "making value." For decades both companies had done an excellent job of creating products—mainly different types of photographic film and the materials needed to develop that film—that had great value for people and for society as a whole. But over time the value of popular film photography changed. The availability of digital cameras that were more convenient and less expensive to operate than film cameras decreased the value of film, making the traditional core business of both Kodak and Fujifilm steadily less profitable. To remain viable the companies needed to find new ways to make value. As explained, Fujifilm has thus far done that far more successfully than Kodak.

Although it is not yet in widespread use, the concept of *making value* is a particularly effective way of examining the success and failure of individuals, businesses, communities, and nations. Making value is the process of using ingenuity to convert resources into a good, service, or process that contributes additional value for a person or society. While *value creation* is often used to refer to the ability to provide things of worth for the customer or user, *making value* is used here to emphasize the entire system of activities that is necessary to conceive, produce, and deliver these things—especially the design and production processes that often receive less attention in discussions of value creation.

It is important to recognize that it is not only companies that fare better or worse depending on how well they succeed in making value. The welfare of individuals, communities, states, and entire countries depends on their ability to make value and take advantage of that value.

Consider, for example, the development of Research Triangle Park in North Carolina. For six decades, from the 1920s through the 1980s, a healthy textile industry drove much of the state's growth. But by the 1990s, rising living standards in the South combined with greater access to low-cost capacity in other countries led most textile manufacturers in North Carolina and the rest of the United States to relocate overseas. Similarly, tobacco—farming and the production of tobacco products—was a major part of the state's economy through the 1960s, but by the 1990s most of the tobacco industry in the state had relocated overseas.

Instead of stagnating in the wake of these major changes, North Carolina has been replacing its lost jobs in textiles and tobacco with a variety of new jobs in high-tech industries such as analytics, electronics, and pharmaceuticals. In a sense, the state had been preparing for this transition for over half a century, since it established Research Triangle Park (RTP) in the 1950s to foster innovation in the region. RTP is a prominent example of local government working with private industry and academia to create a local ecosystem for innovation that attracts and nurtures high-tech companies and takes advantage of the talented students graduating from nearby universities, including Duke University, the University of North Carolina at Chapel Hill, and North Carolina State University. Encompassing Raleigh, Durham, and Chapel Hill, RTP is now home to a rapidly growing number of companies that make electronics components, design software, and develop nanomanufacturing techniques, while the western part of the state manufactures a large percentage of the world's fiber-optic cables and contains a large number of data centers, including those run by Google, Apple, Facebook, and AT&T. In the face of the decline of its textile and tobacco industries, North Carolina found many new ways to make value.

The examples of Kodak, Fujifilm, and North Carolina illustrate that technologies and global forces, such as expanding access to international markets and workers, change and transform the value associated with a product, service, region, or set of skills. Individuals, companies, communities, and countries that do not change effectively in response can be left behind.

This is a particularly important lesson for Americans to keep in mind today because the economy is facing a number of disruptive changes. Economists and engineers predict, for example, that advances of automation and business processes in manufacturing and across the economy will continue to reshape the labor market in dramatic ways. Computers and streamlined operations are likely to replace an increasing number of workers in a variety of occupations. As much as 50 percent of US jobs are at risk. As this transformation of the labor market continues, the effects on society could be severe unless new types of jobs are created to replace the ones that have been displaced.

Furthermore, globalization and the development of emerging economies have intensified competition. Although US-based businesses remain among the best in the world along multiple indicators of research and output of manufactured goods and services, competitors from several countries are catching up quickly. Because of this many observers worry that, without efforts to strengthen innovation, US-based businesses will not keep pace with the global economy.

As this report discusses, the same forces that are causing these disruptions—technological advances, reorganized business processes, and shifts in the balance of growth throughout the global economy—are also opening up new and exciting opportunities for value creation. There are already many indicators of such positive developments, such as the high demand for computer programmers and emerging innovations and market opportunities. However, a growing number of experts are concerned that, unless actions are taken to get in front of these changes, they could have significant consequences for the future prosperity of the United States and for individuals, companies, and other organizations in the country. Just as these changes are not caused by any one economic sector or set of actors, they also create a need—and therein a significant opportunity—for many actors to respond to these challenges. The individuals, companies, and countries that can truly understand these changes and act on them will be the ones that are most able to prosper in the 21st century.

Introduction

American business and government leaders have drawn considerable attention to the need to strengthen US manufacturing to support innovation and job creation, especially as employment in manufacturing has declined steadily for more than a decade. But framing the debate on manufacturing in terms of whether it is done in the United States or overseas, or how to repatriate repetitive production jobs, misses the big picture. The issue is one of *making value* as opposed to *making things*. The way products and services are conceived, designed, made, and distributed is changing, and the way the middle class is supported will likely change as well. Actions aimed at strengthening American innovation and job creation need to account for these changes and focus on maximizing value along manufacturing value chains.

Manufacturers depend on complex networks of activities that are necessary to deliver their products to market and ensure their utility throughout the product lifecycle. These activities span multiple locations, companies, and economic sectors. They include services and software production to meet and exceed increasingly sophisticated customer needs and desires. And as competition grows, so does the role of these offerings in creating value.

DIVERSITY OF VALUE CHAIN ACTIVITIES AND CONTRIBUTORS

The *value chain*, or *value network*, refers to the system of activities that businesses and workers perform to create a product, deliver it to market, and support it until the end of its life (Porter 1998). These activities include research and development (R&D), design, production, supply chain management, distribution, marketing, and services. Many manufacturing value chains involve a complex network of businesses that create intermediate outputs such as raw materials, components, assembled products, services, and software, with information and iterations flowing back and forth across businesses and stages of the value chain. The value chain for Apple's iPad, for example, involves material and component suppliers (e.g., Corning, Alcoa, Samsung), original design manufacturers (e.g., Foxconn), network providers (AT&T, Verizon), and a wide variety of content creators including large corporations such as TimeWarner as well as individual users (Figure I-1).

Manufacturing value chains traverse the classic divisions of the economy into raw materials production, fabrication of goods, and provision of services and information. Several of these functions may be performed within a single company or even a single establishment. Ford Motor Company, for example, carries out its own R&D, engineering and testing services, and repair and maintenance in addition to manufacturing (Table I-1).

While companies have always been involved in activities that cross economic sectors, it is increasingly difficult to meaningfully categorize companies along manufacturing value chains as providing mainly goods or services. Many companies that traditionally focused on producing and selling goods have developed service-type business models. For example, Rolls-Royce offers a "power by the hour" service that lets customers rent the use of a jet engine rather than purchasing one. Rolls-Royce retains ownership of the engines, monitors their real-time performance, and manages their maintenance and replacement. Such service content has grown in importance among manufacturers. Deloitte Research (2006) found that the fraction of manufacturers' revenues generated by services has grown to approximately 20 percent in the medical device, industrial product, and telecommunication equipment industries and as high as 37 percent in automotive and 47 percent in aerospace. Service content is even more pronounced among many of the world's largest manufacturers, whose main offering is defined by after-sale services.

At the same time, companies primarily known for software and online services have branched into designing and producing manufactured goods. Amazon, for example, established a hardware team that developed the Kindle e-reader and Fire TV digital media player, and is developing a smartphone to more effectively deliver its offerings to customers. Google is partnering with contract manufacturers to produce wearable technology products such as Google Glass.

The interconnected activities along these value chains constitute a large portion of the US economy. Value chains that rely on manufactured goods as either a part of or a necessary means to deliver their main offering (as in the case of software) account for 25 percent of employment, over 40 percent of gross domestic product, and almost 80 percent of R&D spending in the United States.[1]

---

[1] Data from the National Science Foundation, "Worldwide R&D paid for by the company and performed by the company and others"; and the Bureau of Economic Analysis, "GDP by industry" and "Full-time equivalent employees by industry" (available at www.bea.gov; accessed July 28, 2014). Data include goods manufacturing, transportation, warehousing, wholesale and retail trade, computer systems design, Internet and print publishing, broadcasting and telecommunications, and scientific and technical services.

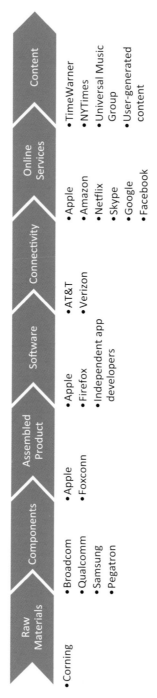

FIGURE I-1 Companies involved in the iPad value chain (not comprehensive).

TABLE I-1 Economic sector classifications of manufacturers' establishments

Company	Economic sector	Primary activity[a]	Location
General Electric	Manufacturing	Aircraft engine and engine parts manufacturing	Cincinnati, OH
	Professional, scientific, and technical services	Research and development in the physical, engineering, and life sciences	Bohemia, NY
		Marketing consulting services	Scottsdale, AZ
	Repair and maintenance	Appliance repair and maintenance	Atlanta, GA
Ford Motor Company	Manufacturing	Automobile manufacturing	Dearborn, MI
	Wholesale trade	Automobile and other motor vehicle merchant wholesalers	Livonia, MI
	Retail trade	New car dealers	Dearborn, MI
	Professional, scientific, and technical services	Engineering services	Dearborn, MI
		Testing laboratories	Dearborn, MI
		Custom computer programming services	Farmington Hills, MI
		Research and development in the physical, engineering, and life sciences	Dearborn, MI
		Marketing research and public opinion polling	Dearborn, MI
	Repair and maintenance	General automotive repair	Wayne, MI
Procter & Gamble	Manufacturing	Sanitary paper product manufacturing	Cincinnati, OH
	Wholesale trade	Other chemical and allied products merchant wholesalers	Cincinnati, OH
	Professional, scientific, and technical services	Testing laboratories	Cincinnati, OH
		Advertising material distribution services	Cincinnati, OH

[a]As indicated by the North American Industry Classification System (NAICS).

THE COMMITTEE'S APPROACH TO ITS CHARGE

The committee was charged with examining the concept of making value for the United States by identifying best practices along the holistic manufacturing value chain—from product development and design to after-sale services, including educational approaches to prepare the workforce and public and private actions to create an effective environment for value creation. Recognizing the complexity of modern manufacturing value chains and the interconnection of the production of goods, services, and software, the committee took a broad approach. The members examined a variety of value chains associated with manufacturing, including those of high-tech services and software that depend on manufactured devices to operate. They also considered activities—such as R&D, professional and financial services, and retail trade—that are needed to bring a product to market and use it throughout its lifetime.

To explore the value chains linked to the manufacturing sector, which comprises the activities and employment of factories, committee members sought and reviewed a variety of data, including economic statistics on output, employment, patents, and other indicators of activities along the value chain. However, by design, the committee was charged with investigating activities that do not easily correspond to the standard organization of economic statistics. When data were available for activities specific to value chains that encompass manufacturing, they were included. In many cases, though, it was either not possible or impractical to gather data on certain services, such as financial or business services, that are specific to manufacturing-related value chains separate from other parts of the economy; in these cases, the committee examined economywide data.

The committee members also gathered extensive information from expert interviews and published research. During their year-long investigations, they sought input from nearly 100 individuals, including directors of manufacturing operations, research managers, entrepreneurs in the United States and abroad, and current and former policymakers at the local and federal levels. They also reviewed literature and interviewed researchers recognized as authorities in productivity, management practices, team dynamics for innovation, and the labor market.

STRUCTURE OF THE REPORT

This report presents the committee's findings and recommendations. Chapter 1 discusses three important factors transforming value chains linked to the manufacturing sector: globalization, advances in computing power and robotics, and improved business processes. These driving factors are changing the nature of work and the types of skills that are demanded on the factory floor, in front-end and sales offices, and in many other areas. Chapter 2 explores opportunities to improve operations and create new product and service offerings by taking advantage of emerging digital technologies and distributed tools. The country's capacity to pursue these opportunities and adapt to transforming value chains will be enhanced by improvements in five areas: widespread adoption of best practices, a well-prepared and innovative workforce, local innovation networks to support startups and new products, capital investment flows, and infrastructure, all of which are described in Chapter 3. Chapter 4 presents the committee's recommendations of specific actions for businesses and federal, state, and local governments and agencies to address these five areas. The Appendix presents a review of factors and challenges that underscore the importance of creating an environment in the United States that continuously attracts and creates businesses and jobs.

REFERENCES

Deloitte Research. 2006. The Service Revolution in Global Manufacturing Industries. Washington: Deloitte.

Porter ME. 1998. The Competitive Advantage: Creating and Sustaining Superior Performance, 2nd ed. New York: Free Press.

1

The Manufacturing Value
Chain in Transition

G lobalization, advances in computing power and robotics, and processes
that improve efficiency and lead time have transformed the way things
are conceived, designed, and made. These developments have in turn
enabled huge improvements in productivity and made a large selection of goods
available for lower costs. US companies continue to capture value throughout
the manufacturing value chain, but manufacturing employment in the United
States has declined. All these forces are causing increased pressures—on both
companies and workers in the United States—that demand increased agility.

GLOBALIZATION

Perhaps the defining feature of the US and world economy over the past several
decades has been globalization. The interconnection of economies around the
world has fueled global trade, investment, technology, and knowledge flows
that have profoundly influenced manufacturing value chains. While there has
always been trade and investment between countries, the current global economic interconnectedness is unprecedented—and becomes more pronounced
with each passing year. Globalization has increased competition as companies
from around the world contend in the same markets as US-based companies.
But it has also allowed companies to distribute activities along the value chain
in locations across the world in search of efficiencies and profit. And it has
allowed US companies to expand into new markets.

*While US-based businesses as a group remain the world leader along
multiple indicators of research and production of high-tech manufactured
goods and services, competitors from emerging economies are advancing
rapidly.*

Increased Competition

Increased trade across national borders and the rise of multinational corporations around the world have increased the competition facing US-based businesses. Although as a group they remain the world leader along multiple indicators of research and production of high-tech manufactured goods and services,[1] competitors from emerging economies are advancing rapidly.

In the global economy, Chinese-based businesses lead the world in total output of manufactured goods, with $2.3 trillion compared to $1.8 trillion from US-based businesses.[2] In high-tech manufacturing—aircraft, spacecraft, communication products, computers, pharmaceuticals, semiconductors, and technical instruments—US-based businesses lead the world (Figure 1-1). But competitors in developing countries, most prominently China, are rapidly increasing their output. As a result, the share of global value added from US-based businesses dropped from 34 percent in 2002 to 27 percent in 2012.

Competition has also increased in high-tech services. US-based companies account for the largest share of global value added in business, financial, and communication services; companies based in the European Union contribute the second largest share (Figure 1-2). However, the US share fell from approximately 44 percent in 2001 to 32 percent in 2012 as the output from developing countries, and certain developed countries (South Korea, Taiwan, Canada, and Australia), increased at a faster rate (NSB 2014).

Global Distribution of Value Chain Activities

The rapidly diminishing importance of geographical boundaries in production and trade affects the manufacturing and high-tech services environment in a number of ways. Businesses are becoming increasingly multinational, no longer located in or predominantly serving one country. They compete for customers around the globe and locate their facilities in countries that make the most business sense.

As a result of these trends, manufacturing value chains have become more widely distributed around the world. Between 1995 and 2009, the cross-border flows of intermediate goods and services as well as final products associated with manufacturing value chains significantly increased (Baldwin and Lopez-Gonzalez 2013). The global distribution of value chains is particularly prominent for final goods such as automobiles, consumer electronics, and pharmaceuticals. Around the world, 40 percent of final goods produced are destined for export. The globalization of intermediate goods such as fabricated metals has also increased but to a smaller extent; 16 percent of intermediate

---

[1] A full description of these indicators is provided in the Appendix.
[2] Data from the World Bank, World Development Indicators. Available at http://data.worldbank.org/indicator/NV.IND.MANF.CD (accessed November 29, 2014).

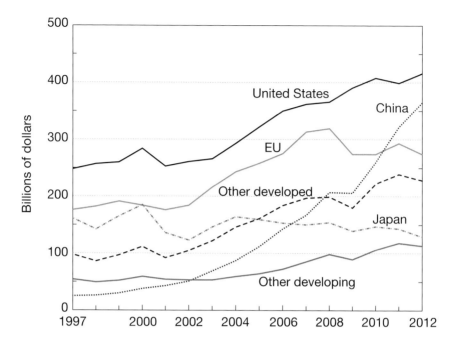

FIGURE 1-1 Output of high-tech manufacturing industries for selected countries, 1997–2012. Source: NSB (2014).

Notes: EU = European Union. Output is on a value-added basis. Value added is the amount contributed by a country, firm, or other entity to the value of a good or service and excludes purchases of domestic and imported materials and inputs. High-tech manufacturing industries are classified by the Organization for Economic Cooperation and Development and include aircraft and spacecraft, communications, computers, pharmaceuticals, semiconductors, and testing, measuring, and control instruments. EU excludes Cyprus, Estonia, Latvia, Lithuania, Luxembourg, Malta, and Slovenia. China includes Hong Kong. Developed countries are classified as high-income countries by the World Bank. Developing countries are classified as upper- and lower-middle-income countries and low-income countries by the World Bank.

goods are now exported for the next stage of production (Baldwin and Lopez-Gonzalez 2013).

Factors Influencing the Location of Facilities

Decisions about where to locate plants and other facilities depend on the factors that are most important to the industry in question. Following are some of the factors that may play a role in the decision of companies to locate various facilities in the United States or in other countries.

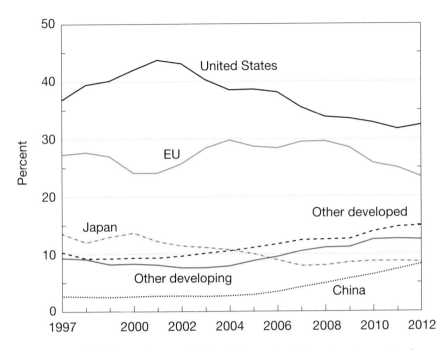

FIGURE 1-2 Global share of commercial knowledge- and technology-intensive services for selected countries, 1997–2012. Source: NSB (2014).
Notes: EU = European Union. Output of knowledge- and technology-intensive industries is on
a value-added basis. Value added is the amount contributed by a country, firm, or other entity to
the value of a good or service and excludes purchases of domestic and imported materials and
inputs. EU excludes Cyprus, Estonia, Latvia, Lithuania, Luxembourg, Malta, and Slovenia. China
includes Hong Kong. Developed countries are classified as high-income countries by the World
Bank. Developing countries are classified as upper- and lower-middle-income countries and low-
income countries by the World Bank.

- *Proximity to markets:* Location of a plant near its markets can have
  a variety of benefits. If the products are expensive to transport (e.g.,
  because they are heavy, bulky, or fragile), it makes sense to minimize
  transportation costs by locating near buyers; such products include
  industrial machinery and household appliances. Similarly, when just-
  in-time delivery is important, locating a plant as close as possible to
  customers is one way to improve delivery times. For manufacturers
  of food and beverage products, being near markets allows them to
  maintain freshness and minimize transportation times for perishable
  items. Proximity to customers also makes it easier to cater to local
  preferences.

- *Proximity to raw materials:* For companies that work with large quantities of raw materials (e.g., metal refiners and fabricators, petrochemical refiners, chemical and plastics manufacturers, food processors) proximity to those materials is crucial. It allows the companies to reduce transportation costs, take advantage of economies of scale, and, often, ensure access to the materials.

- *Cost, availability, and reliability of energy:* For products that require a great deal of energy to produce, the cost, availability, and reliability of energy can be a major factor in location decisions. Perhaps the most dramatic example is aluminum smelting (the production of aluminum from aluminum oxide), which is done by running a strong electric current through a solution containing the oxide. Smelters tend to be located near reliable and affordable energy sources, such as hydroelectric plants (EIA 2012).

- *Location of supply chains:* Certain industries, such as the automotive industry, have complex supply chains that must be carefully coordinated, so it makes sense to locate plants near suppliers.

- *Access to skilled workers:* Some companies need a variety of types of skilled workers, such as engineers, information technology workers, production workers, and craftsmen. While companies can—and often expect to—offer education and training programs to develop the necessary skills in workers, they prefer a pool of potential workers with enough education and skills that they will need a minimum amount of training before they can begin to contribute.

- *Labor costs:* For some product categories labor costs represent a significant percentage of the overall cost. This is the case, for example, with electronics: the final assembly of high-tech products as well as their support and maintenance after the sale both require significant labor. Countries with relatively low labor costs obviously have an advantage in this area.

- *Government regulations and policies:* The location of certain activities, particularly those seen as important to a nation's competitiveness or economic security, may be subject to government regulations and policies. These may include trade restrictions, corporate tax rates and tax breaks aimed at encouraging capital investments, policies that support local production, and regulations concerning safety, product quality, and environmental quality. An example of the influence of government policies on production decisions can be seen in the recent overbuilding of the solar panel industry in China: government subsidies led to the development of a solar panel production capacity that far exceeded demand, which in turn drove down the price of solar panels and made it difficult for companies outside China to compete in the solar panel market (Plumer 2013).

- *Ability to innovate:* Particularly for value chains with rapid product cycles, in which the company that first brings a product to market has a major competitive advantage, the ability to innovate and develop new products rapidly is a major factor in a company's success. These companies prefer to locate in areas with highly developed technological capabilities, necessary talent, and access to financing for research and development. Here too governments can play a major role by offering R&D incentives or funding, promoting higher education and applied research, and purchasing newly developed technologies.

Industry-Specific Location Considerations for Facilities across the Value Chain

The production of lighter-weight commodities such as fabrics and some chemicals can be done almost anywhere, independent of the location of design and sales activities. It therefore generally migrates to locations with the lowest associated costs for labor, energy, and raw materials.

For heavier products such as appliances and automobiles, as mentioned above, there are advantages to locating production close to market because of lower transportation costs (Manyika et al. 2012). Thus, for example, many of the vehicles sold in the United States are produced in the United States or Mexico. In addition to the location of automotive plants, there are advantages to locating the engineering and industrial designers near customers to understand their needs and wants. Designers in the United States can easily interact with people that purchase US vehicles and use the products themselves. Toyota, among others, has located much of the design and production of its minivans and large pickup trucks in the United States because it is home to many of the customers of these vehicles.[3]

In the case of research-intensive products such as those in the biomedical industry, research activities drive the location of activities in the latter stages of the value chain. The biomedical industry today includes a number of new medical devices, therapies, diagnostics, imaging, and medical genomic services that are highly research intensive.[4] Production, testing, and treatment are best located close to the academic and medical center laboratories where the relevant research is done so that the companies can take advantage of their expertise. Biopharmaceutical companies therefore locate their activities in the regions where there is close and strong collaboration between the biotech industry and academic and clinical research. In particular, much of the production of

---

[3] Remarks of James Bonini at "Making Value for America: A National Conference on Value Creation and Opportunity in the United States," February 27, 2014, Beckman Center of the National Academies, Irvine, Calif.

[4] The biomedical industry is also a highly regulated industry, requiring careful navigation of many regulations. The regulatory environment in different countries is also an important factor influencing location decisions in this industry.

biologics, vaccines, and new therapies such as those based on stem cells is based in the United States and Europe because these highly innovative products require close integration of research, development, testing, and manufacturing. In contrast, the production of many older pharmaceutical products—those based on small molecules, where the technologies to produce them have been established for more than 50–60 years—is largely done in Asia, where production costs are lower. However, this is expected to change in the near future as research collaborations with academia grow in Shanghai and companies invest in facilities there.[5]

For products with short development cycles such as consumer electronics, it is important to be able to produce large quantities in a short period of time, so a significant portion of electronics manufacturing has moved to countries such as China that have the ability to scale up production very quickly. For example, Apple might be working with a nine- or ten-month product development cycle and need a supply of 10 million new iPods available in the month or so before Christmas. The rapid and massive scale-up necessary for such a feat requires a location with a mass of process engineers, tooling engineers, and production workers along with the tools and ability to build a new plant in a couple of months. As a result, Apple evolved from a system where manufacturing was done in the United States with parts made in Japan to manufacturing done in Singapore, Korea, Taiwan, and eventually China. And as more of the manufacturing was located in these places, more of the design was moved there as well. Today, although the design of products such as the Apple iMac and iPhone is still led out of California, the engineering designers there work very closely with companies in other countries that are carrying out a significant percentage of the design work on these products.[6]

In short, as value chain activities are distributed around the world, companies can choose to locate their facilities in the country and region that make the most business sense, all factors considered. The United States has certain advantages, such as a large market, a strong legal system, access to affordable and reliable energy, and highly developed R&D and supply chain capabilities that attract facilities where labor costs are relatively less important. Some of these factors are undergoing significant change.

The last decade has witnessed a dramatic change in the availability of low-cost energy in the United States. This is largely due to technological progress in horizontal drilling and hydraulic fracturing combined with the unique infrastructure and industrial ecosystem available to efficiently and cost-effectively

[5] Remarks of Paul McKenzie at "Making Value for America: A National Conference on Value Creation and Opportunity in the United States," February 27, 2014, Beckman Center of the National Academies, Irvine, Calif.

[6] Remarks of Jon Rubinstein at "Making Value for America: A National Conference on Value Creation and Opportunity in the United States," February 27, 2014, Beckman Center of the National Academies, Irvine, Calif.

extract natural gas from shale formations. Coupled with the abundance of shale resources and the separation of US natural gas prices from global oil prices, the United States has a significant competitive advantage for a potentially long-term stable source of energy and feedstock for the electricity, manufacturing, and petrochemical industries (Krupnick et al. 2013; McCutcheon et al. 2011; MIT Energy Initiative 2012). Analyses by the American Chemistry Council (ACC) and the International Monetary Fund show that the shale gas boom has already reversed the US trade balance for the chemical industry from a $9.4 billion deficit in 2005 to a $3.4 billion surplus in 2013 and could reach a $30 billion surplus by 2018 (ACC 2014; IMF 2014). Detailed examination by the ACC of 97 chemical industry projects suggests remarkable growth and value that will occur for the United States (1) during the 10-year initial capital investment phase, when new equipment is purchased and plants constructed, and (2) as a result of ongoing increased chemical output, made possible by lower natural gas prices and increased availability of ethane. The ACC estimates these investments could create 1.2 million temporary jobs and 537,000 permanent jobs (ACC 2014).

On the other hand, the globalization of R&D and supply chain capabilities may present a challenge for the United States, which historically has had a competitive advantage in these areas. Now, other countries are developing research universities, supply chains, and other parts of their R&D and technological infrastructure. As a result, a number of businesses created in the United States are moving abroad to find the resources, supply chains, and capital they need to commercialize and manufacture their products (Box 1-1; Berger 2013). And as countries around the world continue to develop their capabilities, more value chain activities are likely to migrate abroad. If the United States wants to both retain and attract facilities along the manufacturing value chain, it needs to create an environment that supports continuous development of its innovation, manufacturing, and lifecycle services capabilities.

Emerging Markets as an Opportunity for US Growth

Discussions of the standing of US-based value chains in the global economy frequently focus on the challenges presented by other countries, especially a rapidly developing country such as China. With this perspective it is easy to fall into thinking of the situation as a zero-sum game: other countries "winning" the innovation game will inevitably mean that the United States is "losing." But the greatest threat to American prosperity is not that other countries will get better and catch up to—or surpass—the United States; it is that the United States will fail to keep improving itself and thus fall behind as other countries continue to improve.

Charles Kenny, a senior fellow at the Center for Global Development, makes essentially this point in his book, *The Upside of Down: Why the Rise*

BOX 1-1
Do all companies have incentives to carefully consider location decisions?

Inasmuch as companies weigh various factors to determine the best place to manufacture their products, it is natural to suppose that the United States can make itself more attractive to companies. But not all companies go through such a deliberate process when making a location decision, and it is important to understand why and what factors play a role in their decisions.

Consider, for example, FINsix, a company formed in 2010 to build power electronics devices that are smaller, lighter weight, and better performing than traditional power electronics. The company's chief executive officer, Vanessa Green, spoke with the committee at the conference in February 2014 and had follow-up communications with the committee chair.

FINsix's first product, the development of which was funded through a successful Kickstarter campaign, is the Dart, an AC/DC power converter for use with laptop computers that is smaller and lighter than standard laptop converters.[a] The Dart makes it possible to get rid of the "brick" laptop converters that weigh almost as much as some laptops.

At the time of Ms. Green's communications with the committee, the company, headquartered in Menlo Park, California, was on track to deliver the first Darts to Kickstarter backers at the end of 2014. When asked whether the Dart would be manufactured in the United States, she said no and explained that they did not have the time. Companies such as FINsix have an incentive to bring their products to market as quickly as possible to satisfy their funders, and as a result it is difficult to devote much time to the latter stages of the value chain, such as production. FINsix focused mostly on the front end of the value chain, innovating and creating a product that works; once that was done it was time to build, ship, and scale—and begin seeing some income. In short, the selection of another country for manufacturing the Dart seemed to be a default decision: the company did not want to invest the time to determine the best place to make the Dart for different markets.

FINsix is now considering making the Dart in the United States for certain markets, such as laptop converters for the federal government and perhaps for customers who would prefer an American-made product. But it is worth asking what it would take for other US startups to look more closely at the United States for manufacturing their products.

---

[a] "Dart: The world's smallest laptop adapter." Information available at the Kickstarter website, https://www.kickstarter.com/projects/215201435/dart-the-worlds-smallest-laptop-adapter (accessed May 20, 2014).

of the Rest Is Good for the West (Kenny 2014a). There are more benefits than disadvantages to other countries getting wealthier and growing their economies, he argues, even if, in the case of China, a country's economy becomes larger and more powerful than that of the United States. If the United States has only the world's second largest economy, for example, the dollar might lose its dominance as the currency that central banks prefer for their reserves, which could increase US borrowing costs. But Kenny contends that any negative consequences will be outweighed by the benefits of other countries getting wealthier. For instance, US companies will have larger export markets, which will allow them to hire more workers and increase their profits. And other countries' advances in innovating and making value will improve lives in the United States:

> The rest of the world is also inventing more stuff, from modular building techniques in China to new drug therapies and low-water cement-manufacturing processes in India to mobile banking applications in Kenya. We can benefit from those inventions as much as we already benefit from foreign innovators coming to the United States. Among the patents awarded in 2011 to teams at the 10 most innovative American universities, for example, three-quarters involved a foreign-born researcher, according to the Partnership for a New American Economy. As more people in developing countries go to college and as more firms there research and develop new products, there's a potential for increased innovation in both the West and the Rest. That could bring faster progress in a number of different areas here at home, from connectivity to health. (Kenny 2014b)

Perhaps the best way to think of the challenge facing the United States is not in terms of competing with other countries to be the very best in the world, although there are certainly consequences to falling too far behind. Instead, the challenge is to find an approach to strengthen innovation and value creation in the United States, recognizing that the development of other countries can serve as a positive force to achieve this goal.

A particularly important fact to keep in mind is that as the rest of the world—especially the emerging economies—continues to develop, there will be a steadily rising demand for innovative goods and services. This rising demand represents a major opportunity for any US company, large or small, that has the vision and the capability to take advantage of it.

A 2012 report by the consulting firm McKinsey and Company projects that annual consumption in the world's developing countries will reach $30 trillion by 2025 and characterizes this development as "the biggest growth opportunity in the history of capitalism" (Atsmon et al. 2012). The report also predicts that consumption in emerging markets will by then account for approximately half of the world's total consumption, versus just 32 percent in 2010; that 60 percent of the world's 1 billion households with total earnings of at least $20,000 per year will be in developing countries; and that the overwhelming majority of sales of certain types of products, such as electronics and major appliances, will take place in emerging markets. Accordingly, the report predicts that the preferences of consumers in emerging markets will shape a great deal of the world's innovation:

> As e-commerce and mobile-payment systems spread to even the most remote hamlets, emerging consumers are shaping, not just participating in, the digital revolution and leapfrogging developed-market norms, creating new champions like Baidu, M-Pesa, and Tencent. The preferences of emerging-market consumers also will drive global innovation in product design, manufacturing, distribution channels, and supply chain management, to name just a few areas. (Atsmon et al. 2012, p. 7)

The authors make the point that these trends indicate crucial opportunities that policymakers and others who influence US business and innovation will ignore at their peril.

The importance to the United States of making value for the rest of the world is nothing new, of course. For example, another McKinsey study discussed the importance of trade in creating jobs: The increase in US exports between 2000 and 2009 supported 2 million more jobs in the US workforce than would have existed without the exports (Roxburgh et al. 2012). These jobs were mainly in knowledge-intensive sectors, such as high-tech manufacturing and business services and their suppliers. Since 2009, US exports have continued to grow, from $1.58 billion to $2.2 billion in 2012, supporting an additional 1.3 million jobs (Figure 1-3).

Metropolitan areas in particular have benefited from attention to the export market. According to a 2013 report from the Brookings Institution and JPMorgan Chase, exports were crucial to postrecession growth in the country's 100 largest metropolitan areas (McDearman et al. 2013). Between 2009 and 2012, exports were responsible for an average 54 percent of the growth in output in those areas, versus only 37 percent for the country as a whole. Growth in export intensity was also correlated with growth in overall economic output. Between 2003 and 2012, the 10 US metropolitan areas with the greatest growth in export intensity saw an average annual growth in economic output of 3 percent, versus 1.7 percent among those with the least growth in export intensity (McDearman et al. 2013).

Despite the clear benefits of paying attention to the export market, however, many US businesses fail to take advantage of the opportunities. There are a variety of reasons for this failure. A 2009 OECD study identified the major barriers that small and medium-sized businesses face in expanding to foreign markets: shortage of working capital to finance exports, difficulty identifying foreign business opportunities, limited information with which to locate and analyze potential foreign markets, and inability to contact potential overseas

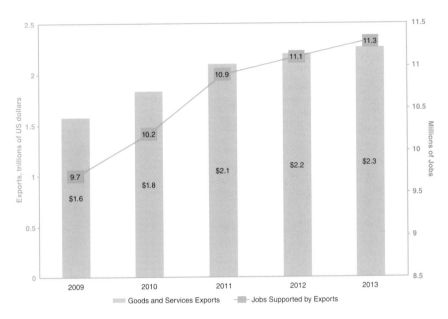

FIGURE 1-3 US exports and estimated jobs supported by exports. Sources: Foreign Trade Division, US Census Bureau; Johnson and Rasmussen (2014).

customers (OECD 2009). Many of these barriers are simply the result of a lack of information and connections to international partners.

In short, emerging markets will offer tremendous potential for US companies in the coming decades, but only if companies and policymakers recognize the potential and then act to develop and maintain the capabilities to take advantage of it.

ADVANCES IN COMPUTING POWER AND AUTOMATION

The ongoing digital revolution is a second major factor driving changes in the US and global economy. Characterized by continual growth in computing power, rapidly improving communication and analytical capabilities, and advances in robotics and control systems, this revolution has had reverberations in every sector of the economy, from the use of global positioning systems on farms and in factories to computer graphics in the entertainment industry. But perhaps no segment has been more deeply affected than manufacturing, where developments such as automation (the control of routine processes by mechanical and electrical devices) and computer-aided design, engineering, and production have dramatically increased productivity and efficiency, reduced lead time, and improved responsiveness to customer needs and preferences.

*Technological advances in computing and automation are allowing com-
panies to increase productivity and efficiency, reduce lead times, and
gain customer insights in their manufacturing, design, and development
activities along the value chain.*

Automobile manufacturing offers a good example of this transformation. Gary Cowger, former president of General Motors (GM) North America
and group vice president of GM global manufacturing and labor relations,
described the company's experience.[7] In the mid-1960s, he said, a typical GM
assembly plant employed about 5,000 workers and produced around 120,000
cars a year. Today, the same sort of assembly plant employs only about 1,500
people to produce the same number of vehicles, but the vehicles are of much
higher quality and have more—and much more sophisticated—individual components. The difference, he explained, lies in the usefulness and effectiveness
of the manufacturing technology used in the plant.

Greater computing power led to more automation in the assembly plants.
At first the automation saved little time because the company was using
machines to put together vehicles that had been designed for manual assembly, but over time the company began designing its vehicles with automated
assembly in mind, and the automation realized its potential as a mechanism for
lowering costs, decreasing production time, and increasing quality.

Implementation of technology throughout the value chain continued, with
the development of higher-speed computers, much more advanced sensor technologies, and numerous materials advances. The result: with more advanced
manufacturing systems and an integrated approach to designing, manufacturing,
and assembling vehicles, it now takes only 30 percent as many people to run an
automobile manufacturing plant as it took in 1965, while the quality, sophisti-
cation, timely delivery, and variety of vehicles have all dramatically improved.

Besides improving the quality of design and manufacturing directly, new
computing tools provide entirely new ways for engineers to get feedback from
customers. For example, Keith Diefenderfer, principal technology director in
the advanced technology center at Rockwell Collins, described the use of virtual
reality immersion labs at the company to let customers try out design variations
before a design is finalized.[8] These labs provide a three-dimensional virtual
experience of a new technology, making it possible for individuals to "use" the

---

[7] Remarks of Gary Cowger at "Workshop on Making Value: Integrating Manufacturing, Design,
and Innovation to Thrive in the Changing Global Economy," June 11, 2012, Venable LLC Conference Center, Washington, DC.

[8] Presentation by Keith Diefenderfer at meeting of the Foundational Committee on Best Practices
of the Making Value for America Study, December 3, 2013, National Academies Keck Center,
Washington, DC.

technology before it is built and offer their insights into any issues or problems with the design. Rockwell Collins can make changes to the virtual environment and then refine a design in response to a customer's comments in as little as an hour. It is a modern approach to prototyping that both increases the chances of meeting a customer's needs and improves productivity by reducing the time and effort needed to create physical prototypes.[9]

A number of other companies are also using immersion labs. A recent study reported that 14 percent of companies surveyed were using this tool and found it highly effective (Booz & Company 2014). For example, Caterpillar, the construction and mining equipment maker, brings in customers as well as assembly line workers and service technicians to test virtual machines and provide feedback on aspects such as usability, ease of manufacturing, and serviceability (Jaruzelski et al. 2013).

Technological advances in computing and automation are allowing companies to increase productivity and efficiency, reduce lead times, and gain customer insights in their manufacturing, design, and development activities along the value chain.

The US automotive industry realized that new technologies were not paying off as expected in terms of improved productivity and began to accept that changes in the processes used in design, testing, manufacturing, and assembly were necessary to achieve the full potential of the new technologies.

IMPROVED PROCESSES

A third factor that has been transforming industry in the United States and around the world is the development and application of new organizational processes, such as lean manufacturing and design for manufacturability that improve productivity and decrease lead time. A number of these processes started in the automotive industry and then spread to other industries, such as aerospace and electronics.

Much of the impetus for new processes over the past few decades has come from the development of new technologies, which required different processes in order to realize their full benefits. As GM's Cowger noted, although the automotive industry began adopting new computer-driven technologies in the 1970s, it was not until the mid-1980s that there were significant changes in the operation of the company's assembly lines. The industry realized that the new technologies were not paying off as expected in terms of improved

---

[9] Ibid.

productivity and began to accept that a greater degree of integration on the enterprise level was necessary to achieve improvements in efficiency. This in turn required changes in the processes used in the design, testing, manufacturing, and assembly of vehicles.[10]

One such change was the growing emphasis on design for manufacturability—the idea that engineers should pay more attention to manufacturing considerations when designing new products. To do this, engineers needed to work more closely with both the people who were assembling the products and the engineers who designed the assembly equipment. No longer could product engineers simply hand off their designs to the manufacturing engineers and forget about them; similarly, the manufacturing engineers could no longer pass their designs to the workers on the factory floor and expect them to take care of things from there. Instead, people began thinking about the interplay between the different components of the process—and transformed the way GM designed its products. It became clear that the upfront design, engineering, and development of a vehicle account for only about 5 percent of the vehicle's total lifecycle cost but the decisions made during this stage determine nearly 75 percent of the lifecycle cost. The company thus began paying more attention to "designing for manufacturability," and engineers and factory workers began communicating more with each other.

Adoption of Lean Production

An even more profound change was the transition from mass production to lean production. In 1990 three researchers from the Massachusetts Institute of Technology published *The Machine That Changed the World*, which described in detail the lean production system developed by the Toyota Motor Company and documented its advantages over the system of mass production that GM and most other automotive companies had been using since Henry Ford popularized the assembly line for constructing cars (Womack et al. 1990). The book quickly led many large companies, in the automotive and other industries, to reconsider their organizational goals and processes. Cowger reported that people began to think of "Big M Manufacturing"—manufacturing in terms of the entire system instead of just along the assembly line—and came to recognize that the problems they had experienced in making new technologies pay off were not just technology problems but also management problems. Since then the lean production system, with its emphasis on a company working collaboratively with employees, suppliers, dealers, and customers, has become the gold standard for production systems.

---

[10] Remarks of Gary Cowger at "Workshop on Making Value: Integrating Manufacturing, Design, and Innovation to Thrive in the Changing Global Economy," June 11, 2012, Venable LLC Conference Center, Washington, DC.

Toyota's production system involves three interconnected elements: (1) The organizational culture puts the customer first, recognizes that the company's most important asset is its employees, and expects continuous improvement. (2) Processes allow operations to be both extremely efficient and focused on quality and problem solving. An important principle of the processes is "just in time," meaning that people and things move through the operations continuously, at the exact time they are needed, with very little waiting. A second, related principle calls for instant attention to production problems: there is an immediate loud signal and a problem solver or team of problem solvers works very quickly to diagnose the problem, contain it, and determine what can be done to prevent it from happening again. (3) Management focuses on developing employees to identify and solve problems to improve operations, and provides the training and incentives for employees to work together as a team to accomplish shared goals.

The practices and principles of Toyota's production system were adapted and applied to many other industries to improve competitiveness. In aerospace, companies such as Boeing and Airbus are constantly looking for ways to improve their operations by streamlining their supply chains and enhancing the efficiency of their factories. Boeing, for example, measures and tracks the day-to-day performance of teams and individuals, looking for weaknesses and ways to improve, and expects its suppliers to do the same. It encourages self-directed work teams to devise ways to improve their work, and the teams work with their supervisors to implement the suggested changes. That empowerment of workers is one of the main reasons that Boeing was able to improve its production efficiency by 50 percent over the past 10 years.

Impacts of Lean Production

A number of studies have reported advantages to the lean production system in manufacturing. One found that firms that had adopted most of its practices were significantly less energy intensive and more productive (Bloom et al. 2008): companies in the 75th percentile in performance of these practices were 17.4 percent less energy intensive than companies in the 25th percentile. The conclusions were based on data from more than 300 manufacturing firms in the United Kingdom.

According to a more recent study based on data from more than 30,000 manufacturing establishments in the United States, companies that adopted lean production principles were significantly more successful than those that did not, according to a variety of measures of success (Bloom et al. 2013). Companies that adopted practices related to monitoring performance, setting targets, and offering incentives were significantly more productive, more profitable, and more innovative and grew faster than those that did not, and the greater the extent to which a company adopted these practices, the more successful it was. This held true after accounting for other factors, such as the company's industry, education level of its workforce, and age of either the company or the particular establishment. Moving from the 10th percentile to the 90th percentile in the performance of lean production practices is associated with a 12 percent increase in value added per employee, a 9.4 percent rise in productivity, a 6 percent growth in employment, and a 2 percent increase in profitability per sale (Bloom et al. 2013).

Between 2005 and 2010 there was a clear increase in the percentage of firms implementing lean production practices, particularly practices involving data collection and analysis. This trend might be due to more companies adopting modern information technologies, such as enterprise resource planning systems. Because these technologies make it easier and cheaper to collect and process data, companies that use them may find that they smooth the way to adoption of principles of lean production (Bloom et al. 2013).

THE NATURE OF WORK IN TRANSITION

Pushed by the forces of increasing globalization, technological advances, and improved processes, employment across the value chain has changed. The manufacturing sector has become more efficient and productive at a faster rate than the rest of the economy, reducing the demand for production workers. At the same time, the nature of work in jobs across the value chain is changing, shifting the education and skills that are in demand.

Effects on Manufacturing Employment

Globalization of manufacturing and pronounced increases in worker productivity have dramatically affected US manufacturing employment. According to the Bureau of Labor Statistics Current Employment Statistics Survey, total manufacturing employment in the United States dropped from approximately 19 million in 1980 to 11.5 million in 2010 (Figure 1-4). It has increased somewhat since then, to 12.1 million in 2014, largely because of growth in three sectors—transportation equipment, machinery, and fabricated metals—that together accounted for about half a million new manufacturing jobs during that period. But it is difficult to know whether this increase marks the start of a trend or is just a momentary upswing in a longer trend of declining manufacturing employment.

The overall employment decline over the past three decades is due in part to jobs being shipped overseas—as when most US apparel manufacturing moved to India and other lower-wage countries—and in part to the increasing efficiency that allows 1,500 workers to assemble the same number of automobiles that took 5,000 workers in 1965. Of course, the push to increase efficiency has itself been driven in part by growing competition from overseas.

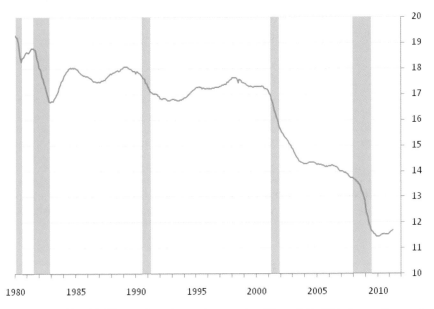

FIGURE 1-4 Monthly US manufacturing employment, in millions, 1980–2011. Source: Bureau of Labor Statistics Current Employment Statistics survey, available at www.bls.gov/ces/. Note: Shading indicates recession.

But enhanced efficiency has also made it possible for some companies to move their manufacturing operations back to the United States, as lean production practices have made them competitive with manufacturers in Korea, China, and Japan (Fishman 2012). These companies are thus able to achieve significant cost savings while creating jobs in the United States (Box 1-2).

> *The number of employees in manufacturing without a high school degree declined from 10 million to less than 2 million in 1960–2010, and manufacturing employment requiring a college or more advanced degree increased by more than 2 million jobs.*

Changing Nature of Work

The decrease in production jobs does not tell the complete story of employment in manufacturing or the larger value chain. Examining employment trends by level of education shows that manufacturing job losses were concentrated in the portion of the workforce without a high school degree (Figure 1-5). Indeed, while manufacturing employment in this part of the workforce declined

BOX 1-2
Return of US production made possible
by lean manufacturing

Over the past two decades, GE Appliances, a $5 billion manufacturing arm of General Electric (GE), had gone through a long period of outsourcing in an effort to remain profitable in the face of global competition and an increasingly commoditized industry. But in 2012 GE began making cutting-edge, high-efficiency water heaters and refrigerators in its Appliance Park plant in Louisville, Kentucky. In a dramatic example of *reshoring*—bringing back to the United States manufacturing that had been offshored—the company opened the first new assembly lines in 55 years in Appliance Park, which at its peak in the 1960s turned out 60,000 appliances a week (Fishman 2012). The committee heard from Kevin Nolan, GE's vice president of technology, who offered some insights into why his company chose to resume making appliances in the United States that, just a few years earlier, it would have made overseas.[a]

GE had been outsourcing its appliance manufacturing to LG and Samsung, Nolan said. It was a good model at first, because the cost of labor was three times lower than in the United States and the company was still able to bring out new products, designed partly by GE engineers and partly through outsourcing. But shipping the products from oversees meant high transportation costs and a lot of cash tied up in inventory. More importantly, the company faced concerns that it would lose engineering and manufacturing skills and that it would become hard to differentiate its products from those of its competitors. When the 2009 recession hit and sales dropped, it became clear the outsourcing model was not sustainable. After trying unsuccessfully to sell the appliance business, GE decided to make a major investment to reshore its manufacturing. It spent $1 billion on new product facilities and manufacturing plants, tearing down existing lines and rebuilding from the ground up so that its new lines are cutting-edge.

In evaluating the reshoring decision, several factors were key, Nolan said. One advantage of reshoring was transportation costs: by locating closer to customers—the new plants were aimed at the US market—the company would see significant cost savings. Shipping a major appliance overseas typically costs about $50 per unit, and it is also expensive to ship from Mexico by rail.

Another important factor was the ability to colocate manufacturing with product engineering and design. GE still had a great deal of its design and engineering capabilities in the United States, and reshoring made it possible to keep them physically near the manufacturing capabilities, especially as GE Appliances is working to create a close connection between the two. Much of the design and engineering of the electronics and software is still done in Korea because the engineers with that expertise are there, but if there are US engineers with equivalent or near-equivalent skills, GE uses them because of the advantages of colocation. Generally speaking, there must be a significant difference in skill sets to do the work in Korea because of the efficiency losses of not being colocated.

However, to take advantage of the lower transportation costs and colocation benefits of manufacturing in the United States, GE needed to dramatically improve the efficiency of its domestic manufacturing for the move to make sense. LG and Samsung have done a much better job of adopting lean manufacturing principles, and they have been doing it for a long time. GE learned that in the United States it generally took about 9 to 10 hours of labor to assemble a refrigerator, in Mexico approximately 10 to 11 hours—and in Korea only about 2 hours. Fortunately, from its work with LG and Samsung, GE had observed their huge efficiency gains from adopting lean manufacturing principles and practices, and it is now working to put those in place at its Appliance Park facilities.

One of the most important principles Nolan pointed to was that labor and management had to work together to develop the manufacturing process. A key factor in the effective implementation of lean manufacturing is the role of operators in problem solving and continuous improvement. Since operators see problems firsthand, they must be empowered to address the problem, with a system in place to support them in solving it quickly. That requires a different way of thinking, Nolan said, and is not easy to accomplish.

Both labor and management had to change their approaches to make the operation a success. Management must recognize the importance of respecting labor and view the production operator as the center of the business. The engineers who design the appliances now recognize that their job is to make it as easy as possible for the operators to do their job.

Labor has also changed its approach, embracing the need to reimagine how manufacturing can be done. In Appliance Park, labor works together with management to improve the efficiency of operations. That is a new approach for labor that is critical to improving efficiency to allow plants in the United States to compete with those anywhere in the world.

On a comparative basis, GE Appliance's reengineered operations are now competitive with the lowest-price producers around the world. And with a continuous focus on production efficiency, their competitiveness continues to improve. Ultimately, GE's bet is that the production line and engineering employees in its new facilities are going to innovate and improve productivity and product features faster than any other company.

Can lean manufacturing lead more companies to bring production back to the United States? Nolan said the potential is great. Companies such as Herman Miller, Caterpillar, and Whirlpool are examples. For heavier products—such as the appliances, furniture, and industrial equipment these companies produce—there is a large incentive to produce them closer to market to reduce transportation costs. For smaller products, manufacturing in the United States will require more automation, Nolan said, to reduce labor costs. But it should be possible to produce heavier products in the United States at lower total costs even without much automation by implementing lean manufacturing practices.

---

[a] Kevin Nolan, interview by Christopher Johnson and Kate S. Whitefoot, May 18, 2014. Since the interview, GE announced the sale of its appliances business to Electrolux for $3.3 billion, following a string of divestments in plastics, insurance, and financial services so as to concentrate on infrastructure and technology products and services. GE credits the reshoring of the appliances division and implementation of lean principles as the reason the business was sold at a high price.

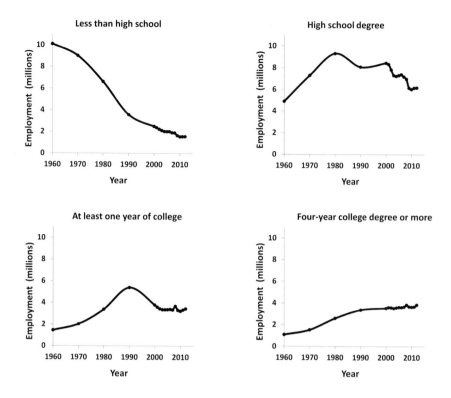

FIGURE 1-5 Trends in employment in the manufacturing sector by level of education, 1960–2012. Data source: Integrated Public Use Microdata Series database (IPUMS-USA; Ruggles et al. 2010).

from 10 million to less than 2 million between 1960 and 2010, manufacturing employment requiring at least a college degree *increased* by more than 2 million jobs (Figure 1-5).

Changing Skill and Education Requirements

Production work in the United States is shifting to require more specialized skills. At Boeing's plant in Everett, Washington, for example, workers control high-tech machines that use indoor GPS and laser-positioning systems to assemble the 787's advanced composite parts, and Boeing expects more automation in its facilities over the next 20 years. There will be increasing emphasis on workers with specialty skills for robotics-controlled maintenance, composites work, precision craftsmanship, computer operations, and radio-frequency identification of parts. Conversely, there will be a significant decline in the subset of manufacturing jobs that consist of repetitive manual skills. Boeing expects that current jobs involving simple tasks such as drilling aluminum and riveting with a bucking bar will continue to decrease and be replaced with higher-skill jobs as new technologies come online.

The shift in the skills needed for production jobs is indicative of a larger transformation across all aspects of the value chain and all sectors of the US economy. In particular, less-educated workers are likely to see declining job prospects and lower wages. Workers without an associate's degree already face much higher unemployment rates—in 2013, 7.5 percent of workers with a high school diploma but no college degree and 11 percent of high school dropouts were unemployed compared to an average of 6.1 percent (Figure 1-6). Wages of men without a college degree have declined 11 percent since 1980 and those of male high school dropouts declined 22 percent, whereas wages of men with a college or advanced degree increased between 20 percent and 56 percent, with the largest gains among those with advanced degrees (Autor 2014). The contrast was less extreme for women, but real earnings growth for women without at least some college education was relatively modest (Figure 1-7).

Many middle-skill jobs are repetitive and procedural and therefore comparatively easy to automate, whereas workers whose jobs depend on manual tasks, such as truck drivers and home health aides, have been more challenging to automate. But this is likely to change in the near future.

Impacts of Automation

The workers who have been hit the hardest are those in so-called middle-skill jobs such as production, operator, clerical, and sales positions, while those in lower-skill jobs such as personal services have actually seen increases in employment and wages. Several economists have attributed this polarization of jobs to increasing automation (e.g., Autor 2010; Brynjolfsson and McAfee 2011).

Many middle-skill jobs are repetitive and procedural and therefore comparatively easy to automate, whereas jobs that depend on manual tasks, such as those of truck drivers and home health aides, have been more challenging to automate. But this is likely to change in the near future. Advances in computing power, machine learning, and robotics are enabling machines to scan a scene, discover patterns, and manipulate objects, enabling innovations such as Google's self-driving car. These trends suggest that in the next 20 years innovations such as truck-packing robots will start to displace the jobs of workers who perform this and other manual tasks. Unless these workers advance their skills, they are likely to see lower wages and declining job prospects.

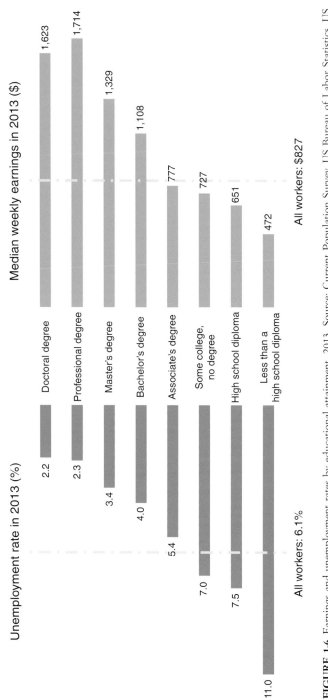

FIGURE 1-6 Earnings and unemployment rates by educational attainment, 2013. Source: Current Population Survey, US Bureau of Labor Statistics, US Department of Labor.

Note: Data are for persons age 25 and over. Earnings are for full-time wage and salary workers.

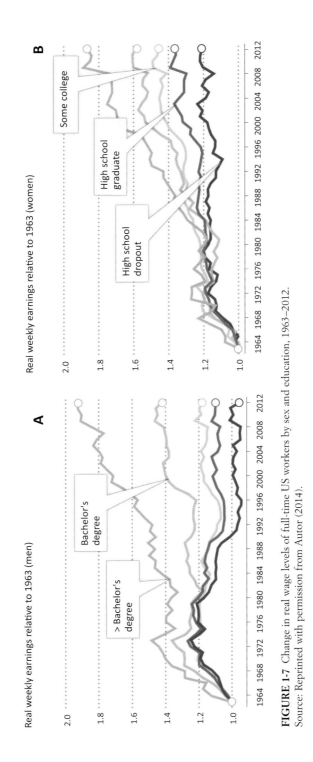

FIGURE 1-7 Change in real wage levels of full-time US workers by sex and education, 1963–2012.
Source: Reprinted with permission from Autor (2014).

The displacement of middle-skill jobs poses serious concerns for the United States. Jobs at this level of pay were important to the growth of the US middle class in the 20th century: many of them were manufacturing jobs, generally at unionized factories, and available to workers without a college degree; they allowed for enough advancement that a worker could stay on a single career path for much of his or her life. The downward pressure on these middle-skill jobs and on lower-skill jobs in areas such as transportation and sales is likely to have a dramatic impact on a large portion of the US workforce. According to one study, almost 50 percent of US jobs are at risk of disruption by advances in computerization (Frey and Osborne 2013).

Growing Skill Mismatches

In addition to—and somewhat related to—the decline of middle-skill jobs, many employers believe there is a growing mismatch between the skills of potential workers and the skills required. An increasing number of jobs—particularly high-wage jobs—require high skills in particular areas, such as engineering and computer programming, and not enough people are acquiring these skills. A Global Talent Management and Rewards Survey of more than 1,600 companies around the world found that 72 percent of them reported that they had trouble finding and hiring critical-skill employees. The situation was only slightly better in the United States, where 61 percent of the responding companies said they had problems recruiting employees with critical skills (Towers Watson 2012). The problem is particularly apparent in the information technology (IT) sector, where as of December 2013 there were some 580,000 vacant positions in the United States (Partovi 2014), and one estimate projected 1.4 million new computing jobs in the United States by 2020 but only 400,000 computer science graduates to fill them (Kuranda 2013).

One major reason that so many jobs are going unfilled is that employers are looking for increasingly specialized capabilities in their new hires. It is not enough, for instance, for an IT worker to be proficient in technical issues; because of the ever more integrated and collaborative nature of jobs and companies, employers would like their IT workers to understand the analytical and business development side of their jobs as well, and such employees are much harder to find than workers who can do just one or the other (Kuranda 2013).

The apparent paradox of employers claiming great difficulty filling engineering jobs and the large number of unemployed engineers in the United States seems to be due to the increasingly specific skills employers expect in their hires (Begley 2005). The result is that workers with the right combination of skills are in high demand and get paid exceptionally well (McBride 2013), while those with skills that do not fit the particular demands of the marketplace often cannot find jobs.

These job market issues, if not resolved, may seriously challenge efforts to create value in the United States.

REFERENCES

ACC [American Chemical Council]. 2014. Keys to Export Growth for the Chemical Sector. Available at www.americanchemistry.com/Policy/Trade/Keys-to-Export-Growth-for-the-Chemical-Sector.pdf (accessed February 2, 2015).

Atsmon Y, Child P, Dobbs R, Narasimhan L. 2012. Winning the $30 trillion decathlon: Going for gold in emerging markets. McKinsey Quarterly. Available at www.mckinsey.com/insights/strategy/winning_the_30_trillion_decathlon_going_for_gold_in_emerging_markets (accessed July 15, 2014).

Autor D. 2010. The Polarization of Job Opportunities in the US Labor Market: Implications for Employment and Earnings. Paper jointly released by the Center for American Progress and the Hamilton Project of the Brookings Institution. Washington.

Autor DH. 2014. Skills, education, and the rise of earnings inequality among the "other 99 percent." Science 344(6186):843–851.

Baldwin R, Lopez-Gonzalez J. 2013. Supply-Chain Trade: A Portrait of Global Patterns and Several Testable Hypotheses. Working Paper 18957. Cambridge, MA: National Bureau of Economic Research.

Begley S. 2005. Behind "shortage" of engineers: Employers grow more choosy. Wall Street Journal, November 16. Available at http://online.wsj.com/news/articles/SB113210508287498432 (accessed April 25, 2014).

Berger S. 2013. Making in America: From Innovation to Market. Cambridge, MA: MIT Press.

Bloom N, Genakos C, Martin R, Sadun R. 2008. Modern management: Good for the environment or just hot air? Working paper 14394. Cambridge, MA: National Bureau of Economic Research. Available at www.nber.org/papers/w14394.pdf (accessed April 7, 2014).

Bloom N, Brynjolfsson E, Foster L, Jarmin R, Saporta-Eksten I, Van Reenen J. 2013. Management in America. CES 13-01. Washington: US Census Bureau, Center for Economic Studies. Available at www2.census.gov/ces/wp/2013/CES-WP-13-01.pdf (accessed April 7, 2014).

Booz & Company. 2014. The Global Innovation 1000: Navigating the Digital Future. Available at www.strategy-business.com/article/00221?pg=all (accessed February 2, 2015).

Brynjolfsson E, McAfee A. 2011. Race Against the Machine: How the Digital Revolution Is Accelerating Innovation, Driving Productivity and Irreversibly Transforming Employment and the Economy. Digital Frontier Press.

EIA [US Energy Information Administration]. 2012. Energy needed to produce aluminum. August 16. Washington: US Department of Energy. Available at www.eia.gov/todayinenergy/detail.cfm?id=7570 (accessed April 4, 2014).

Fishman C. 2012. The insourcing boom. The Atlantic, December. Available at www.theatlantic.com/magazine/archive/2012/12/the-insourcing-boom/309166/ (accessed May 22, 2014).

Frey CB, Osborne MA. 2013. The future of employment: How susceptible are jobs to computerisation? Paper. Oxford Martin School, Oxford University, Oxford, UK.

IMF [International Monetary Fund]. 2014. World Economic Outlook: Legacies, Clouds, Uncertainties. Available at www.imf.org/external/pubs/ft/weo/2014/02/pdf/text.pdf (accessed February 2, 2015).

Jaruzelski B, Loehr J, Holman R. 2013. The Global Innovation 1000: Navigating the digital future. Booz & Company. Available at www.strategyand.pwc.com/media/file/Strategyand_2013-Global-Innovation-1000-Study-Navigating-the-Digital-Future.pdf (accessed April 11, 2014).

Johnson M, Rasmussen C. 2014. Jobs Supported by Exports 2013: An Update. Washington: International Trade Administration.

Kenny C. 2014a. The Upside of Down: Why the Rise of the Rest Is Good for the West. New York: Basic Books.

Kenny C. 2014b. America is No. 2! And that's great news. Washington Post, January 17. Available at www.washingtonpost.com/opinions/america-is-no-2-and-thats-great-news/2014/01/17/09c10f50-7c97-11e3-9556-4a4bf7bcbd84_story.html (accessed January 17, 2014).

Krupnick A, Wong Z, Wong Y. 2013. Sector Effects of the Shale Gas Revolution in the United States. Washington: Resources for the Future. Available at www.rff.org/RFF/Documents/RFF-DP-13-21.pdf (accessed February 2, 2015).

Kuranda S. 2013. Nice work if you can get it: The IT shortage is more serious than you think. CRN, November 4. Available at www.crn.com/news/channel-programs/240163468/nice-work-if-you-can-get-it-the-it-talent-shortage-is-more-serious-than-you-think.htm (accessed April 25, 2014).

Manyika J, Sinclair J, Dobbs R, Strube G, Rassey L, Mischke J, Remes J, Roxburgh C, George K, O'Halloran D, Ramaswamy S. 2012. Manufacturing the Future: The Next Era of Global Growth and Innovation. London: McKenzie Global Institute. Available at www.mckinsey.com/insights/manufacturing/the_future_of_manufacturing (accessed April 4, 2014).

McBride S. 2013. Twitter pays engineer $10 million as Silicon Valley tussles for talent. Reuters, October 13. Available at www.reuters.com/article/2013/10/13/net-us-siliconvalley-engineers-twitter-idUSBRE99C03R20131013 (accessed April 25, 2014).

McCutcheon R, Schlosser P, Misthal B. 2011. Shale gas: A renaissance in US manufacturing? PWC Report. December. Available at www.pwc.com/us/en/industrial-products/publications/shale-gas.jhtml (accessed February 2, 2015).

McDearman B, Donohue R, Marchio N. 2013. Export Nation 2013: US Growth Post-Recession. Global Cities Initiative: A Joint Project of Brookings and JPMorgan Chase. Available at www.brookings.edu/~/media/Research/Files/Reports/2013/09/17%20export%20nation/Export Nation2013Survey.pdf (accessed July 15, 2014).

MIT Energy Initiative. 2012. The Future of Natural Gas. An Interdisciplinary MIT Study. Available at https://mitei.mit.edu/system/files/NaturalGas_Report.pdf (accessed February 2, 2015).

NSB [National Science Board]. 2014. Science and Engineering Indicators 2014. NSB 14-01. Arlington, VA: National Science Foundation.

OECD [Organization for Economic Cooperation and Development]. 2009. Top barriers and drivers to SME internationalization. Report by the OECD Working Party on SMEs and Entrepreneurship. Paris. Available at www.oecd.org/cfe/smes/43357832.pdf (accessed July 15, 2014).

Partovi H. 2014. Testimony before Committee on Science, Space, and Technology; Subcommittee on Research and Technology; US House of Representatives. January 9. Available at http://docs.house.gov/meetings/SY/SY15/20140109/101630/HHRG-113-SY15-Wstate-PartoviH-20140109.pdf (accessed April 25, 2014).

Plumer B. 2013. China may soon stop flooding the world with solar panels. Washington Post, March 23. Available at www.washingtonpost.com/blogs/wonkblog/wp/2013/03/23/china-might-stop-providing-the-world-with-cheap-solar-panels/ (accessed April 5, 2014).

Roxburgh C, Manyika J, Dobbs R, Mischke J. 2012. Trading Myths: Addressing Misconceptions about Trade, Jobs, and Competitiveness. McKinsey Global Institute. Available at www.mckinsey.com/insights/economic_studies/six_myths_about_trade (accessed July 15, 2014).

Ruggles SJ, Alexander T, Genadek K, Goeken R, Schroeder MB, Sobek M. 2010. Integrated Public Use Microdata Series: Version 5.0 [Machine-readable database]. Minneapolis: Minnesota Population Center.

Towers Watson. 2012. The next high-stakes quest: Balancing employer and employee priorities. 2012–2013 Global Talent Management and Rewards Study. Available at www.towerswatson.com/en-US/Insights/IC-Types/Survey-Research-Results/2012/09/2012-Global-Talent-Management-and-Rewards-Study (accessed April 23, 2014).

Womack JP, Jones DT, Roos D. 1990. The Machine That Changed the World: Based on the Massachusetts Institute of Technology 5-Million-Dollar 5-Year Study on the Future of the Automobile. Riverside, NJ: Simon & Schuster.

2

Opportunities for Value Creation Presented by Digital Technologies and Distributed Tools

The 21st century is ushering in digital and distributed tools that will further transform the way value is created and the job opportunities that result. In particular, emerging digital technologies, such as advanced sensors, and distributed capabilities, such as digital manufacturing and crowdfunding, are changing the process of value creation, the types of products and services that can be made, and the types of people who can commercialize them. It is not possible, of course, to predict exactly how these tools and technologies will evolve in coming decades, or what new tools may emerge, but it is clear that digital and distributed tools, broadly speaking, will become even more important to enable value creation. We offer some extrapolations and speculations based on some of the most dramatic changes so far.

Advances in software and data collection have opened up a wealth of possibilities for companies to better understand customers' needs, optimize design and production processes, and discover new market opportunities. The rise of distributed approaches to everything from raising funds to hiring workers is changing the ways businesses operate. These capabilities are allowing companies to integrate systems of hardware and software and leverage data to provide new and improved solutions all along the value chain. They are making it cheaper and easier for entrepreneurs to develop new products and services and bring them to market.

EMERGING TECHNOLOGIES

Advanced Sensors and Cloud Computing

Developments in advanced sensors and cloud computing are allowing companies to collect vast amounts of information to monitor the performance of their products, provide new services to customers, and monitor their supply chains.

Advanced sensors promise to revolutionize a large number of fields. A traditional sensor—such as a thermometer, an accelerometer, or a detector that senses the presence of a particular gas or chemical compound—produces an electrical signal that can be read by an observer or instrument. Today's advanced sensors can be equipped with onboard computing capabilities that enable them to process the signal, carry out diagnostics, and even take intervening actions (e.g., triggering a warning or adjusting the system controls). They often have the capacity to communicate wirelessly, and tend to be both small and inexpensive to manufacture so that many can be used in concert (Spencer et al. 2004).

These advanced sensors are being used to monitor the structural health of buildings, bridges, and aircraft for signs of fatigue or impending failure. The ultimate goal is to create "smart structures" that not only monitor their status and report needed repairs but also, in some situations, make their own adjustments in real time in response to problems (Spencer et al. 2004).

Alongside the emergence of advanced sensors, cloud computing—the use of remote computing resources (typically accessed over the Internet) for computing, digital storage, and software programs—is also enabling "smarter" products and systems. With cloud computing an individual or business is not restricted to onsite computing resources: it is possible to store data and run software on remote computers. Thanks to the economies of scale enabled by sharing computing resources with millions of users and devices through the Internet, cloud computing has begun to provide better performance and more flexibility at a lower cost than can be achieved with a captive computer system.

The existence of large numbers of advanced sensors with computing and wireless communication capabilities raises the possibility of developing networks of thousands or millions of sensors that monitor large systems—such as a factory floor, a city transportation system, or a supply chain—to provide a real-time global picture of what is happening and respond automatically to a situation. At the same time, networking of the computing capabilities of various sensors could provide the systems with tremendous amounts of computing power. According to a report by the National Research Council, "These networked systems of embedded computers…have the potential to change radically the way people interact with their environment by linking together a range of devices and sensors that will allow information to be collected, shared, and processed in unprecedented ways" (NRC 2001, p. 1).

Digital and Additive Manufacturing

Developments such as digital simulation and additive manufacturing are enabling closer integration of design and production. Improved modeling and simulation are used to digitally represent and analyze prototypes, referred to as *digital manufacturing*, which saves time and money on the testing of physical models and specimens. Boeing, for example, has cut in half the amount of wind tunnel testing that it does by using simulations based on computational fluid dynamics.

Additive manufacturing, or 3-D printing (the process of producing parts by depositing and joining layers of material based on a digital model), is reducing prototyping costs by enabling production in smaller runs for lower costs. As a result, it has the potential to dramatically reduce production times and costs for full-scale production (Wohlers Associates 2011). It is also becoming important for the rapid and affordable manufacture of custom tools required for conventional high-rate manufacturing processes, such as casting cores and drill jigs (Cotteleer et al. 2014), the production of which requires considerable time and cost. Producing them with additive manufacturing can both significantly reduce costs and allow products to get to market much faster.

Distributed and Disintermediated Capabilities

New Internet-based capabilities enable new ways to solve problems and run businesses. For example, electronic communication methods distribute functions—such as raising capital, finding workers, creating products, solving problems—over a large number of entities, increasing the participation of people and businesses throughout the value chain. And they allow individuals to carry out these activities without the need for an intermediary institution. These approaches are already changing how companies form and operate, and the changes will accelerate in the coming years.

Crowdsourced Funding

One important use of the emerging distributed and disintermediated capabilities is to raise capital for businesses and projects. Raising capital has always been one of the most difficult hurdles for startup companies or for existing companies with a new product or approach because they have no track record to convince backers of their likely success. Online crowdfunding sites now make it possible for individuals and businesses to seek funding for projects from anyone who would like to provide support. Perhaps the best known of these sites is Kickstarter, a platform for individuals seeking donations for creative projects; in March 2014 it announced that it had surpassed $1 billion in funding, provided by 5.7 million people.[1] A number of other crowdfunding sites—Tilt, Crowdfunder, Somolend, appbackr, AngelList, and Quirky—have been created to raise funds for businesses.

There are two basic types of crowdfunding sites. One type (e.g., Kickstarter) relies on donations, although many recipients promise something in return for the donation—a copy of the resulting record or digital album in the case

---

[1] "OMG. On March 3, 2014, Kickstarter passed $1 billion in pledges." Kickstarter website, at www.kickstarter.com/1billion?ref=promo&ref=PromoNewsletterMar0314 (accessed April 21, 2014).

of funding for a music project, for example. These sites have funded new businesses developing consumer and technology products, designers offering custom fashion items, retail stores seeking to expand or open new locations, and many more projects. The appbackr site also relies on donations for, as its name implies, the development of mobile apps. The second model is "investment crowdfunding," in which funds are provided either as loans or to acquire an equity position in the company. Somolend, for instance, provides loans to businesses that may have difficulty acquiring funding in more traditional ways. AngelList uses crowdfunding to provide angel investing to tech startups. And Quirky brings together people interested in invention to provide the funding and other support necessary to bring inventions to market (Barnett 2013).

In 2012 crowdfunding raised $2.7 billion for more than 1 million projects worldwide (the vast majority of the funding was for projects in the United States), and the total for 2013 was projected to be $5.1 billion—an increase of 81 percent (Massolution 2013). Thanks to this broad new array of potential funding sources, inventors and startup companies have many options for securing funding. It is still too early to know how widespread this phenomenon will become, but it illustrates the sorts of changes that the Internet is ushering in.

Uses of Social Media

In addition to funding, crowdsourcing has been used to provide businesses with various products and capabilities. Many businesses are using "social media" (a broad and somewhat amorphous term that refers to the group of technologies that enable people to interact remotely, generally in virtual communities or networks) to connect with their users and generate ideas for new innovations. The biotech company Genentech, for example, is using social media to collect information about the experiences of patients receiving cancer treatments. The information helps patients learn more about clinical trials while allowing the company to better integrate patient insights into its decision making for developing new medicines. It is likely that other uses of social media will emerge to assist companies in developing new products and services that people value.

Another example of an approach to crowdsourcing is TopCoder, a cloud-based community of computer programmers that companies can call on to produce software and algorithms for use in computer programs. Companies bring a programming problem to TopCoder, which puts it in the form of an open competition. Winning designs are licensed to the companies by TopCoder, which in turn pays royalties to the individuals who developed the designs. TopCoder also offers regular competitions in which programmers are given specific tasks to solve in a fixed period of time. These competitions allow programmers to sharpen their skills and allow TopCoder to identify talented programmers from around the world, and often the resulting programming solutions are valuable for use in various types of software. In April 2014 TopCoder reported on its website (www.topcoder.com) that more than 600,000 people worldwide had registered at the site (although only a fraction had actually submitted programs or algorithms). TopCoder can be thought of as a global programming marketplace where companies find computer programmers to help them develop software and algorithms and computer programmers find jobs and career development.

oDesk, a third example of a novel use of distributed capabilities, connects businesses with freelance workers from around the world. It is the best known of a growing group of "online staffing platforms" where "contingent workers, contractors, freelancers can offer their skills and services for limited projects or even on-going assignments and where organizations and individuals can post their requirements or put tasks/projects out to bid" (Karpie 2012). oDesk provides online workspaces and communication channels for contractors and workers and handles payments for work done. Its website lists freelancers available in areas such as Web development, software development, writing and translation, administrative support, design and multimedia, customer service, and sales and marketing.

Cloud Computing Services

The emergence of cloud computing is important not only because it enables large amounts of data collection and analytics, but also because it is dramatically reducing the time and costs needed to scale up an organization's computing resources. A number of companies, such as Amazon, Google, IBM, and several startups, have begun offering cloud computing services that allow individuals and organizations to rent "elastic" computing capacity—meaning that they can scale the capacity up or down in minutes—allowing them to acquire exactly the right amount of computing resources that are needed at any one time. This ability eliminates the need to purchase computing capacity for the maximum requirements needed at any time and radically reduces the time and costs required to scale up. For example, large pharmaceutical companies that want to acquire the computing infrastructure to support the development of a new drug can now gain almost immediate access to a 30,000-core cluster for $10,000, whereas previously it would have cost $5–10 million and taken about six months to build (McKendrick 2011).

Convergence of These Distributed Capabilities

The common thread among the emerging distributed capabilities is a broadening of opportunities along the value chain, both for businesses and entrepreneurs that can take advantage of them and for skilled individuals who can contribute to them. Together, these tools are significantly lowering the barriers to entry for entrepreneurship.

Launching a startup traditionally requires considerable capital and time to search for talent, acquire technology, and establish marketing and promotion. The convergence of crowdsourcing, cloud computing, and social media now allows potential entrepreneurs to easily access these capabilities at very low costs. And with the advent of digital manufacturing tools that improve product design, visualization, and analysis, as well as low-cost prototyping equipment such as 3-D printers, hardware startups can also build and test their product concepts more quickly and cheaply (Bradshaw 2013). As a result, startups such as Babybe, which develops products to improve the rehabilitation of premature babies, are able to reduce total development costs to less than $150,000 (*Economist* 2014).

A NEW SWEET SPOT FOR VALUE CREATION

As illustrated in the examples above, technological advances are giving rise to new opportunities for innovation and value creation. The committee refers to these opportunities as *sweet spots* and believes that many of the most promising such opportunities in the coming decade will arise from the integration of hardware, software, data, and people.

Recognizing that sweet spot opportunities are always a product of time and place—a sweet spot now would not likely have been one 10 years ago, nor is it likely to be one 10 years from now, and a sweet spot for the United States is not likely to be one for Brazil or China—the committee asked, Where can the United States focus its efforts over the next decade or so that will offer the greatest rewards in terms of creating value? The answer depends both on the directions that technology is taking now and on the particular comparative strengths (and weaknesses) of the United States now and in the near future.

Integrating Hardware and Software

One of the most powerful ways to create value in the emerging technological milieu is to integrate hardware and software to create products and services that are much greater than the sum of their parts. A good example of this is Apple's development of the iPod.

The iPod: A Case Study

Introduced in October 2001, the iPod became one of the most successful consumer products in recent history. To create it, Apple combined innovations in hardware and software, integrating Toshiba's 5-gigabyte hard drive, the FireWire serial bus, and software developed specifically for the iPod. The first iPod was simply a personal music player and, as such, just one more in a line of devices that allowed individuals to listen to their music wherever they went, like the Sony Walkman and the Rio PMP300 MP3 player. But the iPod did its job in a way that was far beyond what other portable music players were capable of, and in a few years it completely dominated its market.

The iPod's first and most noticeable advantage was the amount of music it could hold. Where the Sony Walkman played a single CD (or, in earlier versions, cassette tape) and the Rio held about 30 minutes of music, the iPod's five gigabytes of storage were enough for about 1,000 songs. And FireWire made it possible to transfer music onto the iPod much faster than could be done with the Rio, which used the slower USB 1.1. But the real advantages were much broader and more lasting.

Other portable music players eventually caught up with the iPod on storage capacity and transfer speed, but the iPod had other characteristics that were not so easily matched (LePage 2006). The most important were Apple's design of the iPod for ease of use, integration with its computers, and the software for the two. Suddenly it was simple to load music onto the player and organize it. Music transferred from a CD onto one's computer would automatically load onto the iPod as well. Indeed, the music library on the iPod was automatically synced with the music library on the computer just by connecting the two devices. No one had seen anything like this before—it seemed almost magical. In addition, the design of the iPod was revolutionary. Apple had thought carefully about what was most important for listening to music and created a simple, intuitive interface that made it easy for users to hear exactly the music they wanted. The only controls on the first iPod were a click wheel and five buttons for navigating a simple menu that provided access to the music library. To people used to having many options on their music players, it might seem too simple, but the designers had created value by finding a way to connect users with their music in a way that was almost primal.

Yet another key to the iPod's success was the development of the iTunes Music Store. By integrating the music player with a music listening and buying service, Apple amplified the iPod's usefulness considerably. Now users could browse through thousands of songs, find the ones they liked, and easily load them onto the iPod. It was a totally new way of browsing and buying songs, and it changed people's music buying habits.

Over time the iPod evolved, adding features and modifying the ways users controlled it. The hard drive was replaced with flash memory, the maximum amount of memory grew to 160 gigabytes, the ability to store and play videos was added, as were a video camera, a spoken menu, a touch screen, more powerful processors that increased speed and even allowed users to play video games, and so on. And every component was integrated seamlessly, not only in the iPod but across all the systems that interface with it.

By September 2010 some 275 million iPods had been sold.[2] Simply put, Apple created tremendous value by integrating hardware, software, and services to allow people to listen to music and watch videos in a way that was far superior to anything that had come before.

Wearable Tech

Such integration is becoming increasingly common. A more recent example is the Pebble, a smart watch released in 2013. Funded by Kickstarter, the Pebble developers originally sought $100,000 but eventually received more than $10 million, making it the most successful Kickstarter project ever (Kosner 2012; Newman 2012).

The success of the Pebble watch was, again, due to the way it integrated hardware and software in a simple, appealing package that performed a number of tasks that people found useful. The watch has an "e-paper" display, a vibrating motor, a magnetometer, and a three-axis accelerometer, and it connects with iPhone and Android smart phones wirelessly via Bluetooth. Apps can be loaded onto the watch that take advantage of the different pieces of hardware to offer various capabilities, such as displaying emails, providing notifications of incoming calls on the smart phone, keeping track of pace and mileage while running or biking, and even providing a golf range finder that tells a golfer how far it is to a hole on any of 25,000 golf courses worldwide.

As Jon Rubinstein, former CEO of Palm, commented, the value is not so much in the hardware or even the software in such a device, but rather in the integration of the two to create services for people and the ecosystem around the device.[3] He pointed to self-driving cars as another example of value that can be created through such integration. These vehicles will require the development of new hardware and, especially, new software, but their success will ultimately depend on the creation of systems in which cars communicate with each other in a way that smoothes traffic flow and avoids accidents while getting individual travelers to their destinations as quickly and efficiently as possible.

---

[2] Information from the Apple iPod + iTunes timeline, available at www.apple.com/pr/products/ipodhistory/ (accessed April 20, 2014).

[3] Remarks of Jon Rubinstein at "Making Value for America: A National Conference on Value Creation and Opportunity in the United States," February 27, 2014, Beckman Center of the National Academies, Irvine, Calif.

Putting Data to Work

Consumer Products

One way to enhance value with devices that integrate hardware and software is to add large amounts of data to the mix. This is the approach taken, for example, by Fitbit, a company that makes devices that keep track of users' activity throughout the day and communicate that information using smart phones to a website where the activity data are analyzed. The devices integrate hardware (a 3-D accelerometer to measure movement, an altimeter to detect when a user is going up or down stairs, a clock, a display, and Bluetooth to connect wirelessly with a smart phone) with software (programs to calculate distance traveled and calories burned, various apps), but their real value arises from the data they collect. A user's activity data are uploaded to Fitbit's website, where they are analyzed to help the user track progress toward fitness goals, quality of nightly sleep, and even, with additional data, diet and nutritional plans. At least these are the initial uses of the data; as more and more users deposit data on activity patterns and related factors, it will be possible to use the information in new and unplanned ways, which will in turn increase the value of the data.

A less readily apparent example emerged from Google's purchase in January 2014 of Nest Labs, a maker of smart thermostats and smoke detectors, for $3.2 billion. The purchase surprised many observers and had people asking why Google was interested in a company that made thermostats and why it would pay so much for a company whose annual revenues were probably only about $200–300 million (Rogowsky 2014). As Rubinstein pointed out, the answer seems to lie in the fact that these smart thermostats, when hooked up in millions of homes, produce a tremendous amount of data.[4] Indeed, it is probably more accurate to think of Nest not as a manufacturer of thermostats and smoke detectors but as a data collection business. Its thermostats collect a large variety of data—not just a home's temperature but information such as when people enter and leave a room, when the lights in a room are turned on or off, and the pattern of energy use throughout the day (Roose 2014).

Nest Labs cofounder and chief executive officer Tony Fadell previously worked for Rubinstein at Apple, where he was a major figure in the development of both the iPod and the iPhone. In an interview with the *New York Times* he described Nest's philosophy this way:

> We are a company that communicates to you, not just to your building contractors, about what you put in your home and why it's important. It's not just about turning up or down the heat, it's about the other experiences that come with turning up or down the heat—what are we doing about energy, what are we doing about your health and safety. (Hardy 2013)

---

[4] Ibid.

In particular, Fadell said, Nest's smart thermostats and smart smoke detectors are very different from the traditional forms of these devices. They are no longer just for setting the temperature or detecting smoke; the way they collect, analyze, and communicate data opens up a world of new possibilities.

> We came from the world of connected smart phones and apps. We don't just see a thermostat with a better user interface; we see a smartphone that has thermostat functions. That is a very different thing. We don't see a smoke alarm; we see a smartphone with a fire sensor. When you redefine the world that way, it opens it up to many more possibilities. (Hardy 2013)

The fact that Fadell and Nest see the world in terms of data and information, rather than in terms of devices, is likely at least part of the reason why Google—itself an information rather than a device company—valued Nest so highly.

Fadell went on to describe a future in which most household devices are smart and able to communicate with one another:

> Every time I turn on the TV, that's information that someone is home. When the refrigerator door opens, that's another sensor, more information. Before, it was about one little brain and one little sensor, very tightly programmed. Now we have disparate things with an interconnection network, a brain that can evolve and sensor networks that can evolve, all interacting with these learning patterns. (Hardy 2013)

In short, these interconnected systems of hardware, software, and data offer a whole new area in which innovation can grow and evolve, presenting countless opportunities for creating value.

Automotive Industry

In addition to consumer and household goods, other sources are being explored for the use of integrated systems of devices and information. A great deal of data are already being collected and used for automobiles, as James Bonini of Toyota observed at a conference held by the committee. "It is already happening where personal devices are connected to automobiles," he said. "The automobiles will collect a lot of information about how they are used. That will go back to manufacturers. It will go back to dealers." Much of that information will end up in the cloud, where it will be processed and the results will be used for new types of innovation.[5]

---

[5] Remarks of James Bonini at "Making Value for America: A National Conference on Value Creation and Opportunity in the United States," February 27, 2014, Beckman Center of the National Academies, Irvine, Calif.

One car that is already collecting a tremendous amount of data about its use is the Nissan Leaf, which has been described as a "veritable smartphone on wheels" (Svarcas 2012, p. 169). Many of the data collected by the Leaf relate to use of the car's battery and electric functions, such as the battery's charging history, use management, and deterioration, and the functioning of the car's electrical system. But the car can also collect many other data; it keeps track of the use of the headlights and air conditioning, for instance, and the GPS system tracks the vehicle's position, speed, and distance traveled. The Leaf's electronic data recorders (EDRs), the automotive version of the black boxes used in airplanes to provide information in case of a crash, are programmed to begin operation whenever the car's airbags deploy and to record the vehicle's speed, braking or acceleration, and the status of the airbags. The Leaf also has electronic modules that keep track of idling, acceleration, and braking and can be used to provide information about an owner's driving habits (Svarcas 2012). Nissan's warranty for the Leaf requires that customers visit a certified dealer annually to download information about the use of their vehicle and install software updates that improve performance based on aggregated customer data.[6] Experts in the industry expect this sort of data collection and use to become more common.[7]

Another impetus for collecting data from cars is the push to develop self-driving cars. As automation plays an increasing role in the driving of cars—from systems that automatically apply the brakes when a car gets too close to another to full-fledged self-driving systems that require little or no input from the driver other than the destination—more and more data will need to be collected from the vehicles. In the beginning the data will likely be used for such purposes as analyzing drivers' habits in order to develop ways to make cars and driving safer and more efficient, but as more cars become self-driving, roads and highways could become systems of cars interacting through the cloud. This is clearly an area in which there is tremendous value to be created.[8]

Healthcare Value Chains

The amount and types of health-related data are growing rapidly. Clinical trials for drugs and devices generate great volumes of data on responses to treatment, side effects, patient adherence to treatment, dropout rates, and the progress of diseases and other conditions in patients who are given placebos as well as those undergoing treatment. Basic research provides information on a broad

---

[6] According to the 2013 Nissan Leaf Warranty Information Booklet.

[7] Ibid.

[8] Remarks of Jon Rubinstein at "Making Value for America: A National Conference on Value Creation and Opportunity in the United States," February 27, 2014, Beckman Center of the National Academies, Irvine, Calif.

range of scientific and medical issues, from the biochemical pathways affected by different drugs to details on the mechanisms of various diseases. There is a tremendous amount of—mostly untapped—data in medical records in the offices of individual doctors and medical practices, and the growing use of electronic medical records promises to make these data widely accessible once methods have been devised to address privacy concerns.

Other types of health-related data can be collected from less organized sources such as social media and Internet searches on health-related topics. And the growing amount of genetic health-related data is opening up an entirely new front in the understanding, treatment, and prevention of disease. Genetic information is already used to determine which patients are most likely to benefit from treatment with a particular drug—an area known as *genomic medicine*—and as clinical researchers have access to more genomic data to combine with clinical and other health data, more applications will become possible.

In short, the healthcare industry is an area where it should be possible to create tremendous value by integrating hardware, software, and data to create new applications that promote health and the treatment of disease. For example, medicine is likely to become increasingly personalized as doctors and medical researchers become better able to analyze the effectiveness of treatments in terms of the individual characteristics of patients.

Paul McKenzie of Johnson & Johnson offered another example.[9] The pharmaceutical industry, he said, is trying to move away from the simple model where a doctor provides a medical treatment along with instructions to the patient and leaves it up the patient to follow them: "Here is your pill; come back in 30 days and get another pill." The focus is changing to a systems integration view of the patient, care providers, treatments, and their interactions. For instance, there is an effort to develop apps that help people understand when they should take their medicine in an effort in enhance their adherence to a treatment.

It is a matter of understanding where the source of value is. If it is seen as the discovery and manufacture of better drugs, then the effort will be placed there. But if one understands that patient adherence to medication schedules is a major problem—in neuroscience, adherence from 60 percent of patients is considered good, meaning 40 percent are not getting the prescribed treatment—then it becomes clear that value can also be gained by increasing the percentage of patients who adhere to the treatment schedule. What sort of system—of integrated medicine, software, and information services—would be necessary to get adherence rates of near 100 percent? How might apps and

---

[9] Remarks of Paul McKenzie at "Making Value for America: A National Conference on Value Creation and Opportunity in the United States," February 27, 2014, Beckman Center of the National Academies, Irvine, Calif.

social interactions be used to help people make the right choices for themselves? Answers to questions like these can create a great deal of value.[10]

Efforts to create value in the healthcare system in this way will require a very different mindset from the traditional one. Medicine has always been a very reductive field, with the focus on finding the cause of a disease or condition and treating that cause. And the way medical care costs are itemized and reimbursed in this country—with a specific code for each diagnosis and each individual action—only reinforces that approach. But a great deal of value can be created in the United States by integrating across medical records, clinical trial data, medical devices, pharmaceutical goods, apps, and social media to zero in on the best treatments for individual patients more quickly and more accurately and to help those patients (and perhaps their healthcare providers as well) adhere to those treatments.

TAKING ADVANTAGE OF OPPORTUNITIES

Opportunities accrue to the individuals, companies, and countries that have the capabilities to take advantage of them. If the United States is to continue to excel at creating value, it will have to recognize and promote the necessary skills and capabilities most applicable to the emerging sweet spots.

Individuals

Individuals with certain skills will be most able to benefit from the integration of software, hardware, and data. Because software will be a core element of the sorts of integrated products described above, individuals who know how to design software and write code will particularly find their skills in heavy demand. Indeed, there are already shortages of computer engineers who can design apps for mobile devices, among others (Omojola 2013). Similarly, engineers who can design the devices to run this software, and those that can design the sensors and control systems necessary to operate machinery, will have little difficulty finding jobs. And people who know how to deal with large amounts of data—how to organize, analyze, and gain insights from it—will be increasingly in demand, as the amounts of data available for use continue to grow.

But perhaps the individuals who will be best positioned to take advantage of the coming sweet spots will be those who can work at the various intersections. For example, designers of user interfaces, both hardware and software, work at the intersection between technology and user and, to be successful, must understand user psychology as well as the engineering and design aspects of the product.

---

[10] Ibid.

Another skill set likely to be in high demand is machine learning—the ability to design software and systems that can "learn" from the data collected. Software designers familiar with the challenges of dealing with large amounts of data and with the analytical skills necessary to extract meaning from the data likely will be increasingly valuable in coming years. In addition, systems designers—and, more generally, "systems thinkers"—will be needed to find ways to bring together the various interacting pieces in ways that extract the most value from the agglomeration.[11]

Companies

As with individuals, companies with skill sets most suited for dealing with the integration of hardware, software, and data will be the most likely to thrive. They will have the necessary skills among their employees—hardware designers, software engineers, data experts, and so on—and will be able to integrate across these domains, enabling disparate groups of people to work together in bringing integrated products to market. Apple set the standard, and many companies are following in its footsteps—Samsung, Google, Amazon, and Tesla, for example.

But few companies can afford to maintain all the capabilities needed for a major integrated project. Companies that thrive will be those most able to effectively coordinate with and take advantage of the expertise and capabilities of outside sources and actors along the value chain. For example, optimizing the system of pharmaceutical goods, medical devices, services, health monitoring apps, and information flow between doctors and patients to improve health outcomes requires aligning the activities of hundreds of companies. Businesses that are able to manage systems of this scale and coordinate across many stakeholders will have an advantage.

In addition, companies that can take advantage of emerging distributed capabilities will likely have an edge over those that cannot. For example, crowd-sourced data offer a nontraditional but very powerful source of information whose potential is just beginning to be tapped. With hundreds of millions or even billions of individuals posting information on Twitter, Facebook, LinkedIn, Pinterest, Instagram, Tumblr, Flickr, Google Plus, and multiple blogs, discussion boards, and rating and review sites, companies with the vision and know-how to take advantage of this information should be able to create value in ways not yet imagined.

There will also be opportunities to take advantage of more directed and deliberate contributions from individuals, such as the writing of software (an

---

[11] Remarks of Jon Rubinstein at "Making Value for America: A National Conference on Value Creation and Opportunity in the United States," February 27, 2014, Beckman Center of the National Academies, Irvine, Calif.

area that TopCoder is already exploiting) and the development of apps for valuable specialized tasks. As the world becomes increasingly interconnected, companies will be able to create great value by harnessing the creativity of individuals or groups of individuals who may have no connection with the company except a common interest in a subject or in solving a particular problem.

Advantages and Disadvantages for the United States

How well prepared is the United States to take advantage of digital technologies to make value?

Advantages

One advantage that the United States has is that companies located here can generally find all the components needed for an integrated product or system without going out of the country. Paul McKenzie of Johnson & Johnson made this point when he spoke about his company's decisions about where to locate production of biopharmaceutical products.[12] In the case of older small-molecule compounds, where the science is well understood and manufacturing involves well-known processes, these are essentially commodities, so price dominates location decisions and the company looks for lower-cost sites. But the company is also involved in developing products that require a great deal of innovation, such as the use of stem cells to treat retinal degeneration. The process of making these products available to patients requires research and development; production; administration of the product, which involves surgery and must generally be done very soon after production; and then monitoring and feedback. The tight integration of these processes means that colocation takes precedence over cost in the selection of a site. Johnson & Johnson products such as biologics, vaccines, and stem cells require collaboration between research and development, clinical centers, and manufacturing, and so tend to be produced mainly in the United States and Europe.

More generally, McKenzie told the committee, the United States has an advantage in the way companies are able to collaborate with academia in both basic and clinical research. However, he added, at least in the pharmaceutical industry, he sees other parts of the world—China in particular—catching up with the United States in collaborations between academia and industry.

The United States has a similar advantage in the production of cutting-edge digital devices and services that integrate hardware, software, and data. Between academia and industry, the country has all the necessary skills and capabilities

---

[12] Remarks of Paul McKenzie at "Making Value for America: A National Conference on Value Creation and Opportunity in the United States," February 27, 2014, Beckman Center of the National Academies, Irvine, Calif.

to conceptualize, design, and operate these new technologies. Once the design has been tested and finalized, the manufacturing of the hardware may be done overseas at less cost, but the bulk of the value added will be in what the pieces have been designed to do and their integration with the software in a way that allows users to do things they had never been able to do before. As explained above, the value of the first iPod was not in its five-gigabyte hard drive or its FireWire components or any other of its individual components but rather in their integration, with the appropriate software, to create a music listening experience unlike anything anyone had seen (or heard) before.

The advantage of collaboration between academia and industry should also apply in areas such as software development, the design of user experiences with physical and graphical interfaces, new approaches to gaining insights from large amounts of data, and other areas that require thinking and integrating across multiple fields. The United States has the skills and capabilities required, as well as many companies with experience in integrating different components to create a useful, unified whole.

Yet another US advantage is difficult to quantify but nonetheless considered important by many businesses: Americans have a great, and deserved, reputation for being innovative. Innovators in the United States seem to be more creative, or at least more able—and perhaps more encouraged—to follow their creativity. For the past half century, US academics have dominated the recipients of Nobel prizes for physics, chemistry, medicine, and economics (see, for example, Bruner 2011). Similarly, over the past 40 years at least, the technologies that have been most transformative, from the personal computer and the Internet to smart phones and Google, have been mostly developed in the United States.

To the extent that its creative dominance can be maintained, the United States will have an advantage in areas of technological development that involve the creation of value through fundamentally new and creative innovations.

Disadvantages

Yet certain disadvantages hinder the United States' ability to take advantage of the opportunities posed by digital technologies. One notable shortcoming, discussed in more detail in Chapter 3, is the poor condition of US infrastructure for wireless communications. The infrastructure to connect novel products to digital networks and transfer data between them is critical to enable innovation in digital technologies and the benefits they can provide across the value chain. If the United States improves its digital infrastructure, especially in regions that have poor connectivity, it may be able to stimulate additional innovation and productivity gains.

Another concern is whether the US workforce is prepared to take advantage of digital capabilities and further develop them. The advent of open-source software and the emerging development of open-source designs for objects through 3-D printing have created myriad possibilities to develop new products from building blocks that are readily available. Arguably, there are far more opportunities than people who know how to exploit them. Jon Rubinstein, who works with companies such as Amazon, Qualcomm, and many startups, commented that even in Silicon Valley there are not enough people with the right skills to do all of the work that should be done.[13]

With respect to the potential of data collection and analysis, another disadvantage is that, in some cases at least, it is more difficult to share data in the United States than in many other countries. In his comments to the committee, Paul McKenzie of Johnson & Johnson noted that, in large part because of privacy concerns in the United States, it is much easier to access clinical data in places such as China or India than it is in this country. At clinical sites in China every patient has an electronic medical record, which allows collection of clinical data much more quickly and with much less effort than if the records are on paper or in nonstandardized digital form. Although the technology for creating electronic medical records for all patients exists in the United States just as it does in China, there is a lack of collaboration between the healthcare system, industry, and the regulatory agencies. And broad concerns about privacy, and associated legal protections, have hampered such collaboration and prevented the implementation of standardized electronic medical records in the United States.

The result, McKenzie said, is that companies like Johnson & Johnson are more likely to conduct clinical trials in places where access to information in medical records is faster and easier. He explained that it is critical to have access to real-time data on clinical trial patients to be able to change the course of therapy during treatment if necessary, and the United States is at a clear disadvantage in this area.[14]

REFERENCES

Barnett C. 2013. Top 10 crowdfunding sites for fundraising. Forbes, May 8. Available at www.forbes.com/sites/chancebarnett/2013/05/08/top-10-crowdfunding-sites-for-fundraising/ (accessed April 21, 2014).

Bradshaw T. 2013. Help for the makers to make it. Financial Times, August 1.

Bruner J. 2011. American leadership in science, measured in Nobel prizes. Forbes, October 5. Available at www.forbes.com/sites/jonbruner/2011/10/05/nobel-prizes-and-american-leadership-in-science-infographic/ (accessed April 25, 2014).

---

[13] Remarks of Jon Rubinstein at "Making Value for America: A National Conference on Value Creation and Opportunity in the United States," February 27, 2014, Beckman Center of the National Academies, Irvine, Calif.

[14] Remarks of Paul McKenzie at "Making Value for America: A National Conference on Value Creation and Opportunity in the United States," February 27, 2014, Beckman Center of the National Academies, Irvine, Calif.

Cotteleer M, Neier M, Crane J. 2014. 3D Opportunity in tooling: Additive manufacturing shapes the future. New York: Deloitte University Press.

Economist. 2104. Doctoring devices. Technology Quarterly, September 6.

Hardy Q. 2013. Nest's Tony Fadell on smart objects, and the singularity of innovation. New York Times, November 7. Available at bits.blogs.nytimes.com/2013/11/07/nests-tony-fadell-on-smart-objects-and-the-singularity-of-innovation/ (accessed April 22, 2014).

Karpie A. 2012. Online staffing: Still partly cloudy—clearing trend ahead. Mountain View, CA: Staffing Industry Analysts. July 31. Available at www.staffingindustry.com/site/Research-Publications/Blogs/Andrew-Karpie-s-Blog/Online-Staffing-Still-Partly-Cloudy-Clearing-Trend-Ahead (accessed April 23, 2014).

Kosner AW. 2012. Pebble watch for Android and iPhone, the most successful Kickstarter project ever. Forbes, April 15. Available at www.forbes.com/sites/anthonykosner/2012/04/15/pebble-watch-for-iphone-and-android-the-most-successful-kickstarter-project-ever/ (accessed April 21, 2014).

LePage R. 2006. Five reasons the iPod succeeded. Macworld, October 23. Available at www.macworld.com/article/1053505/ipodfive.html (accessed April 20, 2014).

Massolution. 2013. 2013CF: The crowdfunding industry report. Available at http://research.crowdsourcing.org/2013cf-crowdfunding-industry-report (accessed April 22, 2014).

McKendrick J. 2011. How cloud computing is fueling the next startup boom. Forbes, November 1. Available at www.forbes.com/sites/joemckendrick/2011/11/01/cloud-computing-is-fuel-for-the-next-entrepreneurial-boom/ (accessed February 9, 2015).

Newman J. 2012. Pebble smartwatch pre-orders are sold out, $10+ million pledged. Time, May 10. Available at http://techland.time.com/2012/05/10/pebble-smartwatch-pre-orders-sold-out/ (accessed April 21, 2014).

NRC [National Research Council]. 2001. Embedded, Everywhere: A Research Agenda for Networked Systems of Embedded Computers. Washington: National Academy Press.

Omojola A. 2013. The shortage of developer talent is crushing mobile. Forbes, July 15. Available at www.forbes.com/sites/ayoomojola/2013/07/15/the-shortage-of-developer-talent-is-crushing-mobile/ (accessed April 25, 2014).

Rogowsky M. 2014. 5 reasons Nest sold to Google. Forbes, January 14. Available at www.forbes.com/sites/markrogowsky/2014/01/14/5-reasons-nest-sold-to-google/ (accessed April 22, 2014).

Roose K. 2014. Why did Google just buy a thermostat company? New York, January 13. Available at http://nymag.com/daily/intelligencer/2014/01/why-did-google-just-buy-a-thermostat-company.html (accessed April 22, 2014).

Spencer BF Jr, Ruiz-Sandoval ME, Kurata N. 2004. Smart sensing technologies: Opportunities and challenges. Structural Control and Health Monitoring 11(4):349–368. Available at http://sstl.cee.illinois.edu/papers/Smart_Sensing_Technology.pdf (accessed April 17, 2014).

Svarcas F. 2012. Turning a new LEAF: A privacy analysis of CARWINGS electric vehicle data collection and transmission. Santa Clara High Technology Law Journal 29(1):165–197.

Wohlers Associates. 2011. Additive Manufacturing Technology Roadmap for Australia. Fort Collins, CO.

3

An Ecosystem for Creating Value

The challenges presented by increasing competition and the changing nature of work, and the opportunities presented by digital technologies, will require US companies and communities to strengthen their ability to innovate and create value. New prospects will be available to the people and organizations that can identify market opportunities and execute the business models and resources needed to commercialize solutions. Businesses will likely face increasing pressure to improve productivity and speed to market and will need to continue to reinvent their operations. Jobs along the value chain and the broader economy will be disrupted as advances in robotics and software enable machines to perform more complex tasks. In this environment, the best bet to aid the workforce that has been displaced by these changes is to advance their skills and ensure that the United States has an "innovative ecosystem" that can continuously attract and create jobs along the value chain and in all sectors of the economy.

This chapter presents five areas fundamental to an effective *value creation ecosystem* that will support US-based value chains: (1) widespread adoption of best business practices, (2) an innovative workforce, (3) local innovation networks, (4) flow of capital investments, and (5) infrastructure that enables value creation. It concludes with a discussion of federal programs that monitor the condition of various activities in US-based manufacturing and high-tech service value chains.

WIDESPREAD BEST BUSINESS PRACTICES

There are different forms of innovation, from improving processes to produce and deliver existing products to developing radically new products and services. Along this spectrum of innovation, different best practices are needed.

The techniques and approaches used to reduce lead times are quite sophisticated, and the people who know how to do it well have learned by applying the techniques for 25–30 years.

Best Practices for Improved Productivity, Quality, and Lead Time

There is a dramatic difference in performance between the best- and worst-managed companies in the United States. By improving their practices and management systems, underperforming firms can dramatically improve their productivity, quality of output, and lead time.

Larry Burns, former head of research at General Motors (GM), recounted his experience with GM's supplier recognition program. He noted that many of the top suppliers remained the same from year to year, indicating that some companies are consistently, significantly better than most of their competitors.[1] Research supports Burns' observation. Even in narrowly defined industries in the United States—manufacturers of ready-mix concrete, for example, or of chewing gum—the differences in productivity between companies are enormous. An efficient producer can get twice as much output from the same inputs as another in the same industry during the same period (Syverson 2011). The clear implication is that significant industrywide gains in productivity are possible if the least efficient firms improve their efficiency. Closing the gap between the least and most efficient firms is not essential, however; reducing it by even 25 percent would greatly increase efficiency and growth—and value.

What accounts for productivity differences between companies? Research suggests that management practices based on lean manufacturing play an important role. Levels of management skill associated with these practices are strongly correlated with differences in productivity. This might seem an obvious conclusion, but until 10 or 15 years ago there were few systematic data to show that this is indeed the case. Now such data are starting to be collected. For instance, the US Census Bureau has added a management survey to its annual survey of manufacturers, which supplies data on more than 30,000 manufacturing establishments in the United States. The first year of these data showed that 27 percent of the companies had adopted less than half of a group of management practices involving performance monitoring, targets, and incentives. At the other end of the spectrum, just 18 percent of companies had adopted at least 75 percent of those practices (Bloom et al. 2013). Since these practices are central to lean manufacturing (Womack et al. 1990), the extent to which

---

[1] Comments by Larry Burns at meeting of the Foundational Committee on Best Practices for the Making Value for America Study, September 23, 2013, the National Academies, Washington, DC.

a company follows them offers a good measure of its commitment to the lean production system.

Similarly, the World Management Survey (WMS) has been gathering data since 2004 on the same management practices from more than 10,000 companies in 21 countries. The data show that well-managed companies "massively outperform" competitors that are not well managed: "They make more money, grow faster, have far higher stock market values, and survive for longer" (Bloom et al. 2011).

According to the WMS survey, US firms are, on average, the best managed in the world, although they are not much ahead of companies in Japan, Germany, and Sweden. The countries with the lowest scores were Brazil, India, China, and Greece (Bloom and Van Reenen 2010). But the gap between the bottom and top performers in any country trumps the differences between countries. Several companies in India and Brazil perform just as well as the top performers in the United States, and the bottom 15 percent of companies in the United States rank below the average for firms in China or India, two countries that have the most poorly managed firms, on average, among the nations surveyed.

For US companies to maintain—or increase—their competitive advantage, best practices will have to be widely adopted, especially at the most poorly managed companies. This does not presume that all businesses will apply these management practices exactly the same way, but companies of all sizes and in all industries deal with many of the same problems, such as redundancy of work and the need for continuous improvement, that can be addressed by implementing these practices. Since these practices improve a company's performance, it may seem that it would be easy to encourage companies to adopt them, but in reality it is very challenging.

Toyota's James Bonini, who helps other companies apply the Japanese automaker's lean production system to their own production efforts, articulated some of the challenges. First, he said, is blindness about how much operations can improve. Many companies say they need to reduce their costs by 10 percent or they will have to move to another country or shut down. They are convinced that it is impossible to reduce costs this much while operating in the United States. After working with these companies to implement lean manufacturing, Bonini has found that they can reduce costs by as much as 40 or 50 percent.

Second, the system must be learned by doing. The techniques and approaches used to reduce lead times, for example, are quite sophisticated, and the people who know how to do it well have learned by applying these techniques for 25–30 years and thus have developed very high levels of expertise in achieving the seemingly basic principle of "just in time." When Toyota teaches its methods to another company, it focuses on a particular area in which lean production will be implemented. It instructs the people in the company on the conceptual underpinnings of lean production—the philosophy, technical tools, role of management, and so on—but it also picks two or three company employees to work side by side with the people from Toyota in the implementation. Working with the more experienced Toyota representatives, the company workers learn how to put the principles into practice, and can then share their understanding with other company employees. This process of learning by doing in a company can take as much as two years.[2]

Often, best practices are transferred to new companies through partnerships. GM first learned how to implement the Toyota Production System through a joint venture with the New United Motor Manufacturing, Inc. (NUMMI) plant in Fremont, California. Through this cooperative venture, GM managers learned the principles and practices of the system first-hand and after expending considerable effort spread the knowledge throughout the company (Inkpen 2005). The same was true for GE when it outsourced the production of its appliances to LG and Samsung; Kevin Nolan of GE reported to the committee that the company was able to see firsthand how LG and Samsung operated through these partnerships. These companies have embraced lean production, he explained, and have been doing it for a very long time. "In the US, everyone talks about lean manufacturing but very few people actually know how to do it," he said. Once GE workers were familiar with the lean operations at these companies, they understood how to implement them at GE.

Best Practices for New Product and Service Innovation

Developing new solutions requires the ability to identify market opportunities, the creative skills to formulate new business models, and the acumen to implement the resources needed to create and deliver new products and services that customers value. Many experts have proposed best practices for innovation and new product and service delivery (Kelley 2001; Carlson et al. 2006; Brown 2009; Christensen and Raynor 2003). Several of these practices emphasize the need for businesses to improve their abilities to both (1) understand customers and identify their needs and desires, and (2) repeatedly innovate in distinctive ways to gain a competitive advantage. These two capabilities are becoming increasingly important as competition intensifies and the pace of technological change increases.

Larry Burns spoke with the committee about the necessity of these capabilities: "To succeed in today's hyper-competitive world, companies must explicitly design for positive experiences with their products and services, and constantly innovate around every facet of these customer experiences." An iterative process of understanding customer experiences, building and trying out

---

[2] Remarks of James Bonini at "Making Value for America: A National Conference on Value Creation and Opportunity in the United States," February 27, 2014, Beckman Center of the National Academies, Irvine, Calif. See also Manyika et al. (2012).

a prototype, improving the solution, and applying lessons learned to the next innovation are all critical to maintaining a competitive advantage. Companies that follow these principles—such as Apple, Google, and IBM—are considered among the most valuable brands by consumer and financial rankings, Burns observed.[3]

The concept of *design thinking* exemplifies how businesses can gain insights about their customers and explicitly design for positive customer experiences with their products and services (Brown 2009; Martin 2009). A number of businesses have implemented this approach. Kaiser Permanente, with the help of the consulting firm IDEO, used design thinking to improve its customers' ability to get medical treatment (Brown 2008). IDEO's study of Kaiser's patients revealed that they often became annoyed long before they saw a doctor because of poor experiences checking in and waiting in the receiving room before they were able to meet with a medical professional, especially during nurse shift changes. The study found that wait times were particularly long because nurses routinely spent the first 45 minutes of their shift debriefing the departing shift about the status of patients. IDEO developed simple software and new procedures so nurses could input data throughout a shift and call up previous shift-change notes. As a result, the time between a nurse's arrival and first interaction with the patient was cut in half.

Additional sets of best practices emphasize how businesses can innovate in various ways to differentiate their offerings from competitors and give them a competitive advantage. For example, the innovation strategy firm Doblin has identified "Ten Types of Innovation" that businesses have used to gain competitive advantage. The accompanying case studies suggest that companies improve their advantage by combining different modes of innovation along the value chain—in product performance, services, method of delivering offerings to customers, and other areas (Keeley et al. 2013).

Empirical research has provided some support for a number of the principles promoted by the approaches described above. Many studies have tested the driving forces of successful product or process innovation by analyzing the characteristics of large businesses and their propensity for innovation. While these studies have conflicting results regarding the influence of some factors, they largely agree on the importance of differentiation, monitoring of customer needs, and company culture (Becheikh et al. 2006). Companies that adopt a *differentiation strategy*—developing products that meet customer needs in unique ways and are difficult to replicate—tend to innovate intensively and achieve a greater competitive advantage. Monitoring customer behavior to understand the evolution of buyers' needs and desires has been shown to be beneficial for innovation. And businesses that have a CEO who sets challenging goals,

---

[3] Remarks of Larry Burns at meeting of the Foundational Committee on Best Practices for the Making Value for America Study, September 23, 2013, the National Academies, Washington, DC.

employees who are empowered to take on new projects, and a structure that encourages interaction between functional units—in other words, a company culture of innovation—tend to deliver more innovative products and services.

With respect to monitoring customer needs, several methods center on "customer-led" practices, involving customers directly in business decisions and responding quickly to their feedback. These practices may entail customer focus groups, customer interviews, or ethnographic studies of customers using a product. For example, the software company Intuit assembled a 6,000-person "inner circle" of customers to serve as a standing focus group (Allen et al. 2005). A company with even greater customer involvement is Quirky, which solicits ideas from inventors and posts them to its online community for feedback. The ideas with the most positive feedback get turned into prototypes, which are then further reviewed by users, who make suggestions for improvements, packaging, and marketing and also play a role in setting the price (*Economist* 2012). The *lean startup method* applies similar customer-led principles to business startups, to learn from customers and get feedback from the market (Ries 2011).

Exactly which set of best practices is best suited for a particular firm will depend on various factors: whether the company is new or established, its size, location, industry, and so on. Best practices can be identified for widespread use by further investigating businesses with excellent innovation performance across a wide variety of industries and contexts and determining the practices they have in common.

Spreading Best Practices

An important way to improve the US capacity to create value will be to spread recognized best practices to as many companies as possible, making the best a little better and the worst a lot better.

Partnerships are a well-known means of spreading operational best practices such as lean production, perhaps best exemplified by the GM–Toyota NUMMI plant. Several government programs have also developed to help facilitate the transfer of best practices across businesses and industry sectors. For example, the Manufacturing Extension Partnership (MEP) administered by the Department of Commerce collaborates with manufacturers to help them adopt lean production, formulate export plans, and reduce energy use, among other services. These partnerships have successfully spread lean production best practices and are generally considered worthy of investment (NRC 2013c); according to the partnership's website, "since 1988, MEP has worked with nearly 80,000 manufacturers, leading to $88 billion in sales and $14 billion in cost savings, and it has helped create more than 729,000 jobs."[4]

---

[4] More information at http://nist.gov/mep/about/index.cfm (accessed February 5, 2015).

Could comparable partnerships be used to spread best practices for identifying market opportunities and commercializing solutions? A number of biotech and pharmaceutical companies have recently established research partnerships to analyze open databases of clinical and genetic information, with the aim of collaboratively developing new drugs and diagnostics (Lund et al. 2013a). These partnerships benefit the participating companies by expanding involvement in problem solving and testing and reducing licensing and transaction costs when firms can access knowledge produced by the collaborative network (Battelle 2012; David et al. 2010). Experience with similar partnerships shows that networks with private sector leadership and funding are more likely to be associated with higher business outcomes (Kingsley and Klein 1998). Other important considerations are the compatibility of participating businesses in terms of creating a cooperative environment and achieving a critical mass of individuals who are both sufficiently knowledgeable and empowered to make decisions on behalf of their companies (Welch et al. 1997; Huggins 2001).

AN INNOVATIVE WORKFORCE

At the heart of innovation are the innovators themselves—the people who generate new ideas for creating value and who, with help, turn those ideas into reality. The first step in encouraging innovation, therefore, is to ensure a steady supply of innovators.

The most basic incubator of such talent is the education system, which should develop students' skills in science, technology, engineering, and mathematics (STEM) and real-world problem solving from kindergarten through college, continued education, and training in the workplace. Improving the participation of women and people of diverse races and socioeconomic backgrounds in STEM education and hiring is also important to create innovative teams in businesses.

There is a need to give more students access to hands-on experiences designing and making, and to nurture the urge to innovate.

Education and Training

Critical Skills and Experience

Innovation requires scientists, engineers, technicians, operators, managers, analysts, and many others with the skills to conceive of an innovation and then develop it from idea to reality. Perhaps the most fundamental skills for innovation along the value chain are those in the STEM disciplines, from software engineering to tooling operations and from molecular biology to social psychology.

US competitiveness depends on improving STEM education and increasing the number of students who pursue it (NRC 2011). By many accounts, the US system of higher education remains the best in the world. However, a number of concerns persist about US STEM education, particularly K–12 science and math education and the quantity of science and engineering college graduates, and these concerns negatively affect the perception of the country as an attractive place to locate activities along the value chain. (The Appendix discusses these concerns in detail.)

Critical thinking and creativity are as important as technical skills. It is not enough to learn facts and procedures by rote; students need to learn to evaluate a situation by asking questions, observing, collecting further information, and subjecting the collected data to a thoughtful analysis to identify mistakes and weaknesses and come up with alternative possibilities. Creative critical thinkers constantly probe and evolve their own interpretations and ideas.

It is therefore important that schools and other educational programs nurture the urge to innovate. Experience in various STEM programs around the country has demonstrated that opportunities for students to innovate solutions to real-world problems can be an effective way of teaching principles of engineering, science, and mathematics (NRC 2011), the fundamentals of innovation in a hands-on way, and the role of innovation in improving the world around them.

Students are gaining experiences developing real-world solutions through a variety of formal learning and extracurricular programs, but these opportunities are not yet widespread. The Next Generation Science Standards (NGSS) identify engineering design content and practices that all K–12 students should learn (NRC 2013a). The supporting framework for these standards had been adopted by eleven states as of August 2014. In addition, "Maker Spaces," community spaces with the parts and equipment necessary to build mechanical and electronic devices, and programs such as *FIRST* (For Inspiration and Recognition of Science and Technology), which offers design and build competitions for K–12 students, offer students opportunities to create and make real-world products. But many schools and communities do not yet have similar opportunities in place. More students need access to hands-on experiences designing and making things (NRC 2013a).

Some colleges and universities are providing their students with opportunities for such experiences. UC Davis started the Engineering Student Startup Center (ESSC) and the Engineering Fabrication Laboratory (EFL) to provide all undergraduates and graduates with the resources to develop and prototype new ideas and experience what it is like to be an entrepreneur. The extensive ESSC and EFL facilities include a machine shop and a rapid prototyping machine that allows students to 3-D print their designs.[5] Stanford University also has a number of programs to encourage design and real-world experience. For example, Stanford Biodesign involves students and faculty from over 40 departments and provides innovation classes, mentoring, fellowships, and career services.[6]

Access to higher education and training is especially important for lower-skilled workers, who are most affected by technological developments and changing business models.

Higher Education

As the nature of work changes across the value chain, access to higher education and training is especially important for lower-skilled workers, who are most affected by technological developments and changing business models.

Unfortunately, this part of the workforce also faces greater barriers to higher education. The rising costs of college attendance put greater strain on low-income families, and students from these families lack the social supports to help them complete degree programs (Haskins et al. 2009). Only 30 percent of college students from families in the lowest quartile of the income distribution complete their degrees, less than half the completion rate of the average student (Holzer and Dunlop 2013).

Higher education organizations are experimenting with new models to reduce costs and improve access. One basic approach toward accomplishing this goal is to track the cost-effectiveness of university and college programs. Measuring the productivity of these programs, taking account of outcomes and costs, is seen as the most promising strategy for improving the affordability of a quality higher education (NRC 2012b, p. 1). Metrics that show the productivity of university and college programs are needed to enable students to make informed decisions about the value of enrolling in a particular program, and to support decisions about ways to improve cost-effectiveness.

Organizations are also reducing the costs of higher education and improving access by creating more flexible pathways to enter and exit degree programs, particularly with community colleges, which offer low-cost pathways to transfer to bachelor's degree programs. Almost one half of Americans with bachelor's degrees in science or engineering attended a community college at some point

---

[5] Information is available at http://engineering.ucdavis.edu/undergraduate/engineering-student-startup-center/ (accessed February 5, 2015).

[6] Information is available at http://biodesign.stanford.edu/bdn/index.jsp (accessed February 5, 2015).

(Tsapogas 2004). Unfortunately, students, especially from low-income families, often face barriers that prevent them from successfully transferring from community colleges to university programs. These include a lack of advising services to help them choose appropriate courses, insufficient information about the transfer process and financial aid options, and a lack of alignment of curricula content at community colleges with university programs (ACSFA 2008). Several state and national initiatives have sought to reduce these barriers; for example, the National Articulation and Transfer Network and the Kentucky Council on Postsecondary Education have implemented programs to help institutions better align their course requirements, provide students and advisors with more information about transfer guidelines, and offer mentoring services to support the transfer process (ACSFA 2008).

In addition to transfer programs, organizations are establishing methods to recognize knowledge and skills gained without a completed degree (Lund et al. 2013b). The Manufacturing Institute, for example, has developed the Manufacturing Skills Standard Certification System, which recognizes specific production skills applicable to all manufacturing industries. Nationally recognized certification systems also exist in energy and information technology fields. Such certifications allow students without a bachelor's degree to gain higher-paying jobs—on average, workers with certificates earn 20 percent more than high school graduates do (Carnevale et al. 2012). Moreover, some of these credentials are "stackable," meaning that they are part of a series that can be accumulated over time and count toward a degree-granting program. This structure is particularly important for lower-income students and dislocated workers, who often have family and work responsibilities that prevent them from completing a continuous degree program (Ganzglass 2014). Widespread recognition of these types of certifications by degree-granting programs and employers has the potential to significantly improve employment and career outcomes (Ganzglass et al. 2011).

Preparing the workforce with the education and skills needed to succeed in the face of changing technologies and business models requires shared responsibility among educators and employers—and both parties can share the benefits.

Some colleges are experimenting with online education and computer-based tools to reduce costs and boost retention. These tools enable personalized learning and rapid feedback that improve student learning. A study at Carnegie Mellon University found that college students studying statistics through an online environment, supplemented with weekly face-to-face meetings with an instructor, learned a full semester's worth of content in half the time of equivalent classroom instruction (Lovett et al. 2008). Online tools can also reduce education costs (Bakia et al. 2012): once online course materials are developed, they can be reused at a relatively low cost and distributed to large numbers of students. Further cost reductions can be achieved by redesigning courses to allow for more effective use of an instructor's time and transferring some activities to computers.

Preparing the workforce with the education and skills needed to succeed in the face of changing technologies and business models requires shared responsibility among educators and employers—and both parties can share the benefits. Recent partnerships between schools and employers combine classroom-based learning with work experiences, many of which target students who are at higher risk of dropping out of high school and put them on a track to a college degree and a skilled job. For example, several schools in Chicago and New York follow the Pathways in Technology Early College High School (P-TECH) model, in which employers partner with high schools and community colleges to design a curriculum that meets state learning standards and leads students to higher degrees and entry-level jobs in areas such as computer science, biotechnology, electromechanical engineering technology, and robotics. One particularly valuable aspect of P-TECH schools is that they can help disadvantaged students—often members of underrepresented minorities—transition to college and to a well-paying job and career (Dossani 2014). Students attend high school for six years, with the chance to earn an associate's degree along the way. They are paired with mentors from the school's corporate sponsor and given opportunities to participate in summer internships and job shadowing. Graduates gain skills that are attractive to the corporate partner and are given priority consideration for jobs at these companies (see, for example, Foroohar 2014).

Chicago's Austin Polytechnical Academy is another example of an employer-education partnership that leads lower-income students to higher degrees and skilled jobs. The school provides an advanced manufacturing curriculum that was jointly developed with local manufacturers, providing instruction in manufacturing, design, engineering, and business skills such as networking, in addition to standard courses in math, science, English, and social studies. The partnership with employers enables students to participate in internships and job shadowing experiences over the summer and be mentored by experienced professionals, all of which can give them the skills and social supports necessary for a college degree and career.

Nationwide, there are over 660 schools like this in 36 states, Washington, DC, and the US Virgin Islands[7] and they are showing very promising results.

---

[7] Information about these programs is available from the National Academy Foundation (http://naf.org/statistics-and-research; accessed February 5, 2015).

Their students are substantially more likely to complete high school—90 percent receive a diploma, compared to a national average of 78 percent—and are better prepared than their counterparts to earn a higher degree (Webb and Gerwin 2014).

Governments at all levels share responsibility for providing access to high-quality education. But the committee is concerned that local, state, and federal investments in education are not adequate to ensure that all American students have access to an effective, rigorous education. In fact, according to a study from the Center on Budget and Policy Priorities, at least 34 states provided less funding per student in the 2013–14 school year than in 2007–08 (Leachman and Mai 2014), and 13 of those states cut student spending by more than 10 percent during that period.

If the United States is to retain international strength and leadership in value creation, committed support for the education of all its current and future workers must be a priority.

Employer Training Programs

In addition to partnerships with high schools and higher education institutions, employer training for both lower-skilled and professional employees is an important component of advancing workforce skills. Employer training programs raise the earnings potential of low-skilled as well as professional workers and can substantially increase productivity, benefiting both employer and employees (Bartel 1994; Veum 1995; Ichniowski et al. 1995; Krueger and Rouse 1998; Hansson 2007). A review of employers' return on investment from training programs indicates that returns may be much larger than previously believed, in some cases as high as 100–200 percent (Bartel 2000). For example, a team-building training program provided by Garrett Engine to randomly assigned maintenance teams led to faster maintenance response and completion times by the teams that received the training, reducing total downtime by 14 percent. The company calculated the return on investment of the training at 125 percent (Pine and Tingley 1993).

Despite the sizable returns employers can receive from training programs, both employers and employees report that the current level of employee training, especially in small businesses, is not adequate (Lynch 2004). Small businesses in particular face a number of barriers that prevent them from delivering adequate training programs (Lynch 2004; Panagiotakopoulos 2011; Dutta et al. 2012). The costs of training programs per employee are higher in smaller businesses because they cannot spread fixed costs over a large group of employees. Turnover rates are often higher, discouraging employers from investing in the skills of workers lest they leave the company. And small businesses struggle more with the time and short-term productivity losses required for employees to receive training.

Policies and partnerships can address many of the barriers to training. Government-provided training subsidies for employers are one option (Lynch 2004), as are partnerships or mediating organizations that coordinate between employers, labor, and government to provide workforce training. These partnerships entail co-investment by employers, employees, and government; the training curriculum is jointly determined by these three sets of stakeholders; and the skills learned in training are certified to ensure uniform quality standards and portability between employers (Lynch 2004). Examples of these types of partnership training programs geared toward particular sectors—information technology in New York and manufacturing and health care in Milwaukee— have improved job outcomes for employees and low-income adults struggling in the labor market, with almost a 30 percent increase in earnings attributed to the training (Greenstone and Looney 2011).

Teams

The Importance of Teams for Innovation

Despite the popular image of a lone inventor, successful innovation is almost always the result of teams working together on a problem. Innovation requires talented people all along the value chain: engineers, scientists, and business leaders who develop inventions and create jobs; tooling engineers and others who create processes to produce goods faster and more efficiently; marketing and business analysts who gain insights on customer needs and market opportunities; and technical support and retail personnel who deliver positive customer experiences.

Individuals and individualism do play an important role in innovation, particularly in the discovery or inventive stage (Černe et al. 2013; Ramamoorthy et al. 2005). Indeed, nations with more individualistic cultures tend to have more patents and highly cited scientific publications (Taylor and Wilson 2012; Gorodnichenko and Roland 2011; Shane 1993). However, effective teamwork is necessary for successful innovation, and collectivism and collaboration are linked to higher rates of commercialized innovations (Černe et al. 2013; Tiessen 1997).

Thomas Malone, a professor in the Sloan School of Management at the Massachusetts Institute of Technology (MIT) and director of the MIT Center for Collective Intelligence, suggested that in the United States there is a "cultural illusion" about the importance of individuals. There are certainly occasions when individuals make a big difference, but that happens much less often than most people think, he said. In general, success in innovation and other business areas is due to groups of people and how well they work together rather than to the contributions of one or a few singular individuals.[8]

Team Performance

Even when individuals set aside their personal goals and work as a team, putting together an effective team is more complicated than simply assembling competent individuals. There has been a great deal of research into what makes an effective team, but there is still much that remains unknown. Malone and his colleagues have demonstrated a *group intelligence* that is analogous to the IQ of individuals (Woolley et al. 2010): it is not a function of the intelligence of the individual members but rather of the way they interact. In other words, putting the smartest people in a group will not necessarily result in the smartest team.

Experiments conducted by researchers at MIT and Carnegie Mellon University have shown that the average intelligence of group members and the highest intelligence level of any individual in the group are not very good predictors of a group's performance (Woolley et al. 2010). Rather, the research shows that a team's performance is strongly associated with the social and cognitive characteristics of its members: their ability to interpret each other's emotions and to speak in turns rather than dominating discussions is strongly correlated with the group's effective performance on a large variety of tasks such as brainstorming, solving puzzles, building objects with a complicated set of constraints, making moral judgments, and negotiating over limited resources. Cognitive styles are also an important aspect of group intelligence, which increases as cognitive diversity increases—but only to a point; if a group becomes too cognitively diverse, the collective intelligence tends to drop (Aggarwal et al. 2013).

Multiple studies on collective intelligence have found that group performance is strongly associated with the percentage of women, leveling off at about 75–80 percent of the group (Woolley and Malone 2011; Aggarwal et al. 2013). Women tend to score higher on tests of social perceptiveness—the ability to read team members' emotions in their facial expressions and the ability to listen to others. Since these abilities are important factors of collective intelligence, the number of women in a group is a good indicator, on average, of the team's performance. This finding has intriguing implications for the makeup of teams in businesses, and is especially important for manufacturing, high-tech services, and entrepreneurship groups of all types, where women are substantially underrepresented (Khanna 2013; Klobuchar 2013; Mitchell 2011).

---

[8] Remarks of Thomas Malone at "Making Value for America: A National Conference on Value Creation and Opportunity in the United States," February 27, 2014, Beckman Center of the National Academies, Irvine, Calif.

Any effort to increase the nation's ability to create value should have, as one of its core principles, a commitment to making sure that all individuals have an equal opportunity to take part in that effort.

Diversity

Beyond the number of women on a team, it is important to take the overall diversity of a team into account, for both practical and ethical reasons. The practical reason is that greater diversity of thought generally leads to more and better innovation: The more perspectives and life experiences and ways of thinking a team brings to a problem, the more ideas are likely to be generated. Furthermore, teams that include people of different gender, race, cultural or socioeconomic background, sexual orientation, and other characteristics are more likely to produce solutions that will appeal to a broad array of customers.

There is widespread belief among business executives that diversity among employees and managers is a competitive advantage for their company and that diversity is actually a key factor in successful innovation. A survey of more than 300 senior executives worldwide found that 85 percent of them believed that a diverse workforce, offering different perspectives, leads to greater innovativeness (*Forbes* 2011). A growing body of evidence supports the notion that diversity of demographic characteristics, thought, and culture is important for team performance and overall business outcomes (Hong and Page 2004; Hoogendoorn et al. 2013; Johansson 2004; Barta et al. 2012). One study analyzed data from the National Organizations Survey, which samples for-profit businesses across the United States, and found that racially diverse businesses had, on average, greater sales revenue, more customers, and greater market share than businesses that were not racially diverse; the same was true of businesses with a relatively even mix of male and female employees compared to those that were less gender diverse (Herring 2009).

Research has shown that cultural diversity also is linked to innovation performance and economic growth. Regions with more cultural diversity, in terms of the share of their foreign-born population, tend to have higher productivity, R&D output, and entrepreneurship (Niebuhr 2010; Ottaviano and Peri 2006).[9] In fact, over 25 percent of engineering and technology companies in the United States had at least one foreign-born member on their founding team (Wadhwa et al. 2007). The benefits of cultural diversity are also evident in patenting rates (Parrotta et al. 2012; Chellaraj et al. 2005): In 2011, 76 percent of patents from

---

[9] As one might expect, the benefits of cultural diversity are strongest when all team members are fluent in the same language; otherwise, the positive effects of cultural diversity on innovation are counteracted by communication difficulties (Parrotta et al. 2012).

the top ten patent-granting universities had at least one foreign-born inventor (PNAE 2012). By some estimates, increasing the share of immigrant college graduates by 1 percent increases the per capita patenting rate by as much as 18 percent (Hunt and Gauthier-Loiselle 2010).

The ability to attract talented students and workers from diverse cultures around the world has historically been a great strength of the United States. Since 2000 foreign students with temporary visas have earned 39–48 percent of US doctoral degrees in the natural sciences and engineering (NSB 2012). In the past, a large percentage of these foreign nationals remained in the United States after graduation and started new businesses or otherwise contributed to the economy. But there is evidence that many are now choosing to return to their home countries. It also appears that students from the top programs are somewhat more likely to choose their home countries over staying in the United States compared to students from less high-ranking programs (NSB 2012). The loss of this cultural diversity, especially from such highly trained graduates, is not favorable for the US economy.

Apart from the practical benefits of diversity, there is also a clear ethical argument to be made for diversity. Creating value is not an end in itself. It improves life by making it possible for people to have more of the things they need and want. A variety of historical, social, and psychological barriers, including innate biases, have prevented many underrepresented groups from gaining equal access to value creation opportunities (Steele 2010; Kahneman 2011; Banaji and Greenwald 2013). If this portion of the population is deprived of the opportunity to take part in the creation of value, they will also be denied the opportunity to partake of many of its benefits. The people most intimately connected with the creation of value are also those who tend to have the highest-paying jobs and thus the greatest ability to enjoy the benefits of innovation and a growing economy. Moreover, they enjoy not only the financial benefits of value creation but also the psychological benefits of knowing they are contributing value to their fellow human beings. Conversely, individuals who are left out of value creation find themselves not only at the bottom of the socioeconomic ladder but also deprived of the chance to contribute in this important way.

A country that is truly successful in making value will leave none of its citizens behind. Thus any effort to increase the nation's ability to create value should have, as one of its core principles, a commitment to making sure that all individuals have an equal opportunity to take part in that effort.

Some companies have recognized the importance of diversity and have implemented programs that have successfully increased recruitment and retention of women and underrepresented minorities. Deloitte & Touche, for example, instituted an effort to track the progress of women in the company, ensure transparency of mentorship and promotions, and promote better work-life balance for all employees (Harrington and Ladge 2009). Through this initiative, the company was able to close the gap in turnover between women and men and achieve a higher number of women in top positions than at any of its competitors. In the early 1990s Xerox established a goal of becoming the employer of choice for women and minorities. The company created an internship program for women and minorities and revised its hiring and promotion practices to create more lateral promotion opportunities and publicize the criteria for these promotions to all employees (NRC 1994). By 2010, 50 percent of managers at Xerox were women and 23 percent were minorities, up from 23 percent and 19 percent, respectively, in 1991 (Butterfield 1991; Xerox 2013). The percentage of all employees that were women rose from 32 percent to 52 percent over that period and minorities increased from 26 percent to 39 percent.

In addition to corporate initiatives, several universities have implemented successful efforts to improve the recruitment and retention of women and underrepresented minorities. The University of California, Berkeley redesigned its introductory computer science courses and eliminated aspects that studies showed deterred women. It reoriented the courses to emphasize the relevance of computing to real-world problems, beginning each class with a discussion of a recent tech-related news article, and added team exercises. Enrollment of women in introductory computer science classes significantly increased as a result, reaching just over 50 percent—the highest percent in the history of Berkeley's digital records. Although the overall share of female computer science majors at UC Berkeley and Stanford is still only 21 percent, the shift in introductory computer science classes is a good first step in the right direction (Brown 2014). Other exemplary higher education programs include the University of Michigan Women in Science and Engineering Residence Program (WISE-RP), the multi-university Gateway Engineering Education Coalition, and the Meyerhoff Scholars Program at the University of Maryland, Baltimore County (UMBC). All of these programs demonstrate eight characteristics that contribute to their success: (1) institutional leadership, (2) targeted recruitment, (3) engaged faculty, (4) personal attention (such as mentoring for individual students), (5) peer support, (6) enriched research opportunities outside the classroom, (7) bridges to the next level (for example, through connections with industry), and (8) continuous evaluation (BEST 2004).

LOCAL INNOVATION NETWORKS

Innovation does not happen in a vacuum. The old stereotype of a lone inventor working heroically and single-handedly to come up with new creations never was accurate—Thomas Edison had an entire "invention factory" devoted to innovation. In today's increasingly complex and interconnected world, innovation efforts are most likely to be successful in the context of *innovation networks* that connect innovators, investors, customers, workers with appropriate skills and talents, industry, academia, policymakers, regulators, and other stakeholders. Innovators in academia benefit from links to industry, and vice versa, and both benefit from links to policymakers and regulators since government policies and regulations can have a tremendous effect on the prospects of innovations.

Elements of Innovation Networks

Innovators need access to a variety of resources if they are to develop their ideas into marketable products. They need access to low-cost capital, for instance—from investors who are willing to provide funding for projects that carry a certain amount of risk. Such investors are quite different from those who invest in established firms with less perceived risk, and they are not found everywhere.

Innovators need access to people in a broad range of disciplines. When they come up against a problem that requires a particular talent to solve—say, a machine-learning problem—they need someone who has that talent. And because most innovation today is interdisciplinary, innovators generally need people with a wide variety of skills. As Frans Johansson (2004) explained in his book *The Medici Effect*, a great deal of innovation is created by bringing together people with different experiences, competencies, and ideas, enabling the application of concepts or tools from one area to a totally different area, resulting in new insights and inventions.

It is also helpful for innovators to establish links with customers. As discussed above and in Chapter 2, feedback from potential customers is one of the best ways to hone an innovation and maximize its chances for success.

The ideal innovation network has all of these components, the players know what their contribution points are, and there are communication and information flows between the components. Investors should interact with academic contributors, the talent pool should interact with industry, and so on. Most successful innovation networks are local—the majority of the components are within a relatively small region—with connections to the broader global innovation ecosystem. The local network facilitates interaction—innovators do not have to look far afield to find what they need—and the outside connections provide links to resources that may not be available locally.

A highly effective example of an innovation network is Silicon Valley, where the synergy among participants has led to decades of innovation. The development of Silicon Valley in the region between San Francisco and San Jose can be traced to two main factors: the presence of Stanford University, with its graduates in the physical sciences and engineering, and a significant amount of military spending in the area on research and development. Once the innovation network got started with early players such as Hewlett-Packard (founded by two Stanford graduates) and Lockheed (located there because of military tie-ins), success built on success, and increasing numbers of innovators chose the area to pursue their dreams.

Outside of Silicon Valley, a number of other local innovation networks have developed—in the area around Boston, the Research Triangle in North Carolina, the area around Austin, Texas, and the Seattle-Tacoma area in Washington. There are also well-established local innovation networks in Israel and Taiwan, among other places.

These networks are typically in areas with a large number of young people with scientific and technical skills, often associated with one or more upper-tier research universities in the area. There are examples of the formation of these networks in more rural areas such as Mondragon, Spain (MacLeod 1997) and Flanders, Belgium (NRC 2008) as well as the more typical urban setting. In Troy, New York, for example, there is a cluster of gaming companies because students and graduates from Rensselaer Polytechnic Institute (RPI) with an interest in gaming decided to stay in town and follow their passion. But, as Heather Briccetti, CEO of the Business Council of New York State, pointed out, the presence of a university is not enough. In the case of one RPI student who decided to turn his interest in gaming into a company, his innovation was supported by an incubator, which provided him with advisors and people with the necessary expertise to transform his idea into reality—a company that now employs 75 people.[10]

Innovation networks in metro areas can be facilitated by urban development decisions. Urban assets such as public and private spaces for stakeholder collaboration and transportation systems can facilitate the connections that stimulate innovation (Katz and Wagner 2014). Unfortunately, the structure of decision making at the state and local levels can make it difficult to support innovation networks. Christopher Cabaldon, mayor of West Sacramento, explained that coordination of local decisions on housing, transportation, zoning, and other urban development issues is needed to attract a critical mass of stakeholders and facilitate connections between them. But the way governments are organized, each department has its own specific mission that it won't or can't compromise even if it is in the state's or locale's interest to make tradeoffs among various objectives. The only way to change that, Cabaldon said, is to change how decisions are made, from this "functional approach" to a "place-based approach" in which decisions are coordinated to optimize a broad set of outcomes, such as quality of life, environmental sustainability, and economic growth, not just how many people are moved or housed. This coordination can allow state and metro area governments to ensure that the urban resources to enable innovation networks are present in the same location.

Coordinated decision making across metropolitan government silos has been implemented in Chicago and Denver (ICF 2009). In 2005 the Illinois state legislature merged the regional planning and transportation planning agencies

---

[10] Remarks of Heather Briccetti at "Making Value for America: A National Conference on Value Creation and Opportunity in the United States," February 27, 2014, Beckman Center of the National Academies, Irvine, Calif.

in the metropolitan Chicago area and the consolidated agency developed the area's first regional comprehensive plan for land use, transportation, housing, human services, environment, and economic development. Going beyond traditional performance metrics of functional planning agencies, the agency is developing performance indicators focused on issues of quality of life, sustainability, and innovation.[11]

Developing Effective Innovation Networks

The development of a local innovation network requires more than the presence of a university, businesses, and a supportive local government. It requires intentional collaboration and assets to take advantage of the strengths of the local area in a deliberate way.

An example of the purposeful development of a local innovation network can be found in the efforts of New York State to create a network of chip manufacturing companies. It began with a suggestion by IBM, which is headquartered in Armonk, that the state work to attract cutting-edge semiconductor manufacturing capabilities. Because of its high taxes, New York is not generally seen as friendly to businesses, but with IBM's prodding the state decided to try to attract the new business. The state does have a number of assets that make it attractive to companies—excellent universities, good infrastructure, proximity to markets, and a large amount of undeveloped land in the northern part of the state—and by offering subsidies it was able to offset the otherwise high cost of doing business there. The state convinced GlobalFoundries to build a major semiconductor factory in Saratoga County, and it now employs 4,000 people. The partnership between New York State, IBM, and GlobalFoundries, along with other actors such as SEMATECH and RPI, jumpstarted a New York–based innovation network centered on semiconductor chips (NRC 2013b).

In her presentation to the committee, Heather Briccetti offered some lessons about how best to develop innovation networks based on her experience with the development in New York State, citing three necessary components:

(1) The private sector must be involved in identifying where the opportunities lie and in creating local ecosystem value. Governments are generally not good at identifying local strengths and opportunities on their own; once a government starts choosing winners and losers, politics inevitably becomes involved and skews the process.

(2) There must be a strong educational system, both primary and secondary, to both attract over the near term and prepare over the longer term people who can contribute to the innovation network. Recognizing that, IBM has become a partner in a New York State project encourag-

---

[11] More information is available at www.cmap.illinois.gov/about (accessed March 26, 2014).

ing local companies to become involved in K–12 education in order to help fill their workforce needs.

(3) There must be partnerships between the private sector and government at the local level. It is not enough to get involved at the state or national levels: companies must work with local governments on local policies, such as education and infrastructure.

Unfortunately, these components are often lacking in many regions around the United States. In particular, larger companies seldom get involved in government partnerships at the local level.[12]

These key components have been instrumental in the development of effective innovation partnerships across the United States. Successful partnerships tend to be characterized by industry initiation and leadership and public commitments that are limited and defined (NRC 2002). In addition, it is important for these partnerships to have clear objectives, cost sharing arrangements, and sustained evaluations of measurable outcomes to support learning and improvement.

Networking among various innovation stakeholders—entrepreneurs, investors, researchers, federal laboratories, local government actors, and others—is critical to innovation networks. Effective leadership and professional management to facilitate this networking have underpinned the development of innovation networks in the Research Triangle, the Sandia National Laboratories region in New Mexico, and the NASA Research Park in California (NRC 2009). Successful networks have been developed abroad as well by leveraging this type of networking. For example, an initiative spurred by the federal state of Brandenburg, Germany, in 1999 provides young entrepreneurs with a mix of individual face-to-face support by a business advisor, group learning workshops, and experience in a business incubator—and led to the support of over 300 startups by 2009 (OECD 2009).

More recently, the US government has begun investing in a series of institutes for manufacturing innovation with the goal of creating a network of regional manufacturing hubs. These institutes, coordinated by the Advanced Manufacturing National Program Office, serve as a point of private-public collaboration for suppliers, schools, colleges, and other organizations to develop and scale particular manufacturing technologies and processes (EOP 2014).[13] As of January 2015, six institutes have been launched in different regions of the country focusing on additive manufacturing, digital manufacturing and design,

---

[12] Remarks of Heather Briccetti at "Making Value for America: A National Conference on Value Creation and Opportunity in the United States," February 27, 2014, Beckman Center of the National Academies, Irvine, Calif.

[13] The AMNPO (http://manufacturing.gov/amnpo.html) is hosted by the National Institute of Standards and Technology.

lightweight materials, next-generation power electronics, integrated photonics, and advanced composites.

The ultimate goal of a local innovation network is to link companies, investors, academia, workers, and government to work together in supporting the creation of new value.

FLOW OF CAPITAL INVESTMENTS

Innovation requires investment. Companies need funding for research and development, capital investments, marketing, and other costs associated with creating value. Yet the evidence indicates that, although many promising opportunities for value creation are opening up, several factors are preventing corporate and venture capital investments in these ideas.

Corporate Investments

The rate of corporate investments has slowed in recent years. One way to gauge corporate spending on investment is to calculate the value of corporate profits minus current investment. For many decades that number was approximately zero, meaning that, on average, corporate investments were about equal to corporate profits. In the past decade, however, the number rose above zero as corporations spent less on investments relative to their profits. Corporate cash balances have risen to record highs, exceeding $2 trillion in domestic reserves by September 2014 (Carfang 2014).

This situation has led some researchers to wonder whether corporations have run out of ideas to invest in. In *The Great Stagnation*, Tyler Cowen (2011), an economist at George Mason University, argued that for centuries the US economy advanced by taking advantage of "low-hanging fruit"—a continent's worth of land to expand into, the labor and contributions of immigrants, and powerful new technologies such as agricultural machinery, the locomotive, and electrical power. With little low-hanging fruit left to harvest, the United States is now in a decades-long economic stagnation, Cowen says, and future innovation will require a very different approach than sufficed in the past.

But many of the largest US companies argue that there is no shortage of problems to solve or of ideas for solutions to them. IBM, GE, Boeing, Apple, and others have a wealth of ideas for potentially valuable innovations—far more than they actually pursue. Why aren't they pursuing them? Why have corporate investments dropped relative to corporate profits? Why are companies sitting on record amounts of cash?

Chris Johnson from GE Global Research pointed to two factors in particular that can slow very large investments: regulatory risk and preferences for short-term returns, especially in the face of stockholder expectations.[14] The first factor relates to uncertainty about future regulations, such as environmental, tax, and fiscal policies, that affect longer-term transactions. For example, the tax credits businesses receive for qualified research expenses expire every two years and must be renewed by Congress, adding significant uncertainty to long-term research expenditures. Regulatory uncertainties increase the risks of investments, and so businesses tend to hold cash as a precautionary measure (Bates et al. 2009).

The second factor relates to a preference for investments that lead to short-term gains over those that pay off in the longer term. In today's financial market, stockholders demand steadily improving performance each quarter. The current tax structure encourages stockholders to hold their assets at least one year by providing a lower tax rate for these investments, but there are no incentives for holding stocks over longer periods. As a result, managers feel pressure to produce short-term earnings to boost their quarterly financial reports, leading to myopic behavior (Bhojraj and Libby 2005; Stein 1989).

The combination of these factors has led businesses to limit the risk associated with investing in transformational "bets" and thus to refrain from pursuing potentially profitable projects that would produce new factories and new jobs. The companies have the cash and financing to invest in these projects but often choose to focus on incremental improvements and short-term projects instead.

The focus on short-term returns has impacted investments not only in today's products but also in emerging technologies that could lead to entirely new industries. A vivid example of earlier long-term research and development, with a horizon on the scale of decades, was the work done at AT&T Bell Labs that led to the invention of the transistor in 1947—and was the basis for the digital industry that exploded over the past 30 years. There were other examples of forward-looking research after World War II at a number of industrial laboratories—for example, IBM's research labs and Xerox's Palo Alto Research Center—as well as national laboratories. Although businesses are still investing in long-term applied research and development, the commitments may not be robust enough to support the explosion of innovation needed to lead US value creation in the coming decades. Moreover, there are concerns that the shift in industrial R&D investments away from fundamental research, such as the work carried out in Bell Labs that led to unexpected transformative discoveries, in favor of applied research with foreseeable results threatens the United States' technological strength (Narayanamurti 2013; NAE 2005).

---

[14] Comments by Chris Johnson at a meeting of the Foundational Committee on Best Practices for the Making Value for America Study, September 23, 2013, the National Academies, Washington, DC.

Long-Term Decline of New Business Activity

Given concerns that large corporations are underinvesting in long-term research, one might turn to new businesses to look for emerging technologies that could drive innovation in the 21st century. Unfortunately, the rate of new business creation has been in a longstanding decline in the United States and the ability of these businesses to access the capital required to commercialize innovations is in short supply.

Entrepreneurial activity in the United States has been declining for the past 30 years, a worrisome trend for jobs because new businesses (startups) are critical for job creation. Older businesses are a key part of the economy—they employ most Americans and are important contributors to productivity growth—but historically they have tended to eliminate as many jobs as they create (Decker et al. 2012). New businesses, on the other hand, create jobs. In fact, businesses as young as five years or less accounted for all job growth between 1982 and 2011 (Haltiwanger et al. 2013). Although many startups don't survive, among those that do are a small group of very fast growing businesses that account for an outsized portion of the job creation and innovative effort taking place in the economy (Decker et al. 2014).

One way to assess the level of US business creation is to look at the number of startups launched each year with at least one paid employee and compare it to the total number of workers in the United States. In 1980, there were more than 35 startups created for every 10,000 workers. By 2010, the number had been cut in half to only 17 (Lynn and Khan 2012).[15] Another important measure of business creation is the share of businesses in the United States that are younger than five years old. This share declined from almost 50 percent in 1980 to less than 35 percent in 2010 (Haltiwanger et al. 2012). This decline is occurring across the value chain, in manufacturing operations, services, and retail.

Considering the important role of new businesses in creating jobs, the decline of business creation in the United States raises concerns about the pace of job creation. The number of jobs created by businesses less than one year old decreased from 4.1 million in 1994 to 2.5 million in 2010.[16] It is important to note that the causes of this slowdown are not known. While not all possible explanations imply severe consequences for the US economy, research has linked the slowdown

---

[15] These statistics do not include the creation of businesses without any paid employees and have been criticized as inappropriate measures of entrepreneurship because they include individuals who claim self-employment because of a lack of job opportunities (Earle and Sakova 2000). Measures of all new businesses, with and without employees, show that the number established each year has been roughly constant; the Kauffman Index of Entrepreneurial Activity shows that the share of people who established either an employer or nonemployer business has fluctuated around 0.3 percent since 1996 (Fairlie 2014).

[16] Data from the Bureau of Labor Statistics website, Employment Dynamics, Entrepreneurship and the US Economy (www.bls.gov/bdm/entrepreneurship/entrepreneurship.htm; accessed August 12, 2014).

in business creation and, more generally, business dynamism—the process by which businesses continually are born, fail, grow, and contract—to declines in productivity growth, innovation, and employment, especially for younger and less educated workers (Acemoglu et al. 2013; Davis and Haltiwanger 2014). John Haltiwanger and his colleagues (2012), using Census Bureau statistics to analyze the role of startups in US job creation, attribute the slowdown of business dynamism to the combination of a long-run secular decline and a short-term accelerated decline caused by the recent recession. The authors explain the important role of new businesses in job creation and the consequences of allowing the current decline to continue (Haltiwanger et al. 2012, p. 2):

> In 2010, 394,000 startups created 2.3 million jobs (these were not simply establishment openings but new firms whose establishments also were new to the economy). This reflects substantial job creation in a time of anemic overall economic activity. Over the same period from March 2009 to March 2010, the net job creation from all US private sector firms was −1.8 million jobs. Without the contribution of business startups, the net employment loss would have been substantially greater.

These are longer-term trends than the recent economic recession, and they are likely to continue even after economic recovery unless actions are taken to ensure that the United States establishes new ways to make value.

Lacking access to long-term, low-cost capital, many potential startup companies with valuable technologies originating in universities and laboratories cannot bring them to market.

Lack of Capital for New Startups

MIT researchers examined the availability of capital for early and later-stage startups in the United States in a 2014 report, *Production in the Innovation Economy*. They found that entrepreneurs face a critical stage of growth once they are ready to move into the pilot phase and early commercialization, when significant capital investments are needed but not available in the United States. In many cases, strategic partnerships of multinational corporations and foreign governments provide the necessary capital and acquire the startup or pull it overseas. The authors identify this lack of capital in the United States as "the critical juncture where innovations developed in the United States are lost," which hinders the creation of significant downstream activities such as manufacturing (Locke and Wellhausen 2014, p. 10).

Researchers in universities and federal laboratories across the United States face this difficulty in accessing financial support to commercialize their innovations. While it is somewhat easier for researchers at a few major universities (Stanford and MIT, for example) that have good connections to investors to find the financial support to develop their technologies into viable businesses, most do not have access to these resources. Chris Silva of Allied Minds, a company that commercializes discoveries from university and federal laboratories, described the impacts of this reality to the committee.[17]

Lacking long-term capital, many potential startup companies with valuable technologies originating in universities and laboratories cannot bring them to market. Silva estimated that at the 40 universities and 40 federal government labs that Allied Minds works with, there are at least 2,000 inventions a year that are potentially commercializable, but Allied Minds has the resources to help launch only six to ten companies a year. And, unfortunately, it has very few peers in the United States that focus on supporting startups to commercialize these technologies from universities and federal labs (Ford and Nelsen 2013). Similar companies, such as IP Group and Imperial Innovations, exist in the United Kingdom, but nowhere else (Moran 2007).

The lack of capital for researchers and entrepreneurs interested in commercializing a new technology is exacerbated by a transition in venture capital in the United States. During the 1990s venture capital provided much of the funding for the countless startups and early-stage companies that yielded dramatic growth in high-tech innovations in Silicon Valley and other regions of the country. Since then, however, venture capitalists have largely abandoned longer-term areas such as biotechnology in favor of funding businesses that are much further along, have less risk, and go up in value every year, if not every quarter (NVCA 2013).

Capital shortage is particularly damaging for innovation in areas such as energy, biotechnology, and materials science. Companies in these capital-intensive long-term fields require patient capital investments with longer time horizons. It may be eight, ten, even twelve years before they begin to fully realize their value, and with the current emphasis on short-term profits that is simply too long. Thus in biotechnology and energy, for example, the money flowing into early-stage companies has essentially collapsed, creating a hole in the innovation pipeline (Margolis and Kammen 1999; NVCA 2013).

Federal programs such as loan and investment programs in the Department of Commerce's Small Business Administration (SBA) have acted to provide some funding for longer-term research and commercialization projects that may not otherwise be supported by venture capital (NRC 2009). However, while this financing was previously directed primarily at startups, it has shifted to fund

---

[17] Remarks of Chris Silva at "Making Value for America: A National Conference on Value Creation and Opportunity in the United States," February 27, 2014, Beckman Center of the National Academies, Irvine, Calif.

older companies, which are less likely to generate significant growth in employment or sales compared to younger companies (Brash and Gallagher 2008).[18]

To operate efficiently, businesses across the value chain must have access to reliable energy and natural resources, transportation, and communication systems. Increasingly, many businesses also need access to computational and digital resources such as high-performance computing grids and information storage.

INFRASTRUCTURE THAT ENABLES VALUE CREATION

In addition to an innovative workforce, capital, and best practices, creating value requires a suitable infrastructure. Without access to appropriate energy sources, transportation, and reliable communications, any type of business will be at a disadvantage. Poor infrastructure hinders value creation.

Contributions of Infrastructure to Innovation

History has shown that the creation of new infrastructure generally leads to technological disruption and massive innovation. The creation of the Internet is one of the best-known examples from recent decades: it has enabled everything from email and Internet shopping to social media sites and home appliances that can be controlled from a distance. The 19th century saw the development of a nationwide railroad system, and the early 20th the distribution of electricity, telephony, the national highway system, and the availability of clean water, which kept the populace healthy. Thus the construction or upgrading of infrastructure can be an important and effective way to encourage innovation and value creation.

To operate efficiently, businesses across the value chain must have access to reliable energy and natural resources (electricity, water, gas, etc.), transportation (via roads, rail, air, and water), and communication (telecommunications, Internet, etc.). Increasingly, many businesses also need access to computational and digital resources such as high-performance computing grids and information storage.

From an economics perspective, the development of infrastructure is often best planned and paid for by government because of positive externalities—public benefits that accrue to those who did not pay for it. Thus one government policy that is most likely to improve the nation's ability to create value in coming years is to support the development of infrastructure. Government

---

[18] Information is also available from the SBA (www.sba.gov/advo) (accessed March 24, 2014).

should think of infrastructure in broad terms—not just the physical infrastructure but also education to produce more skilled workers and the establishment of networks that encourage communication and linkages among the people and institutions involved in innovation.[19]

The United States has many infrastructure assets that facilitate innovation and value creation. Its research infrastructure—the universities, laboratory facilities, and high-performance computing resources that enable cutting-edge research—is widely considered the best in the world (NRC 2012a). Compared to many other countries, the United States also has plentiful access to energy that is relatively cheap and available almost anywhere in the country, especially with the recent surge in the supply of domestic natural gas from shale deposits (although several other countries are considered to have a more reliable electricity supply) (WEF 2013). However, several areas need significant improvement.

The American Society of Civil Engineers issued a "report card" in 2013 that gave the overall state of US infrastructure a D+, based on poor performance across almost all infrastructure categories covering transportation, water, waste, energy, and schools. Only solid waste management received a grade as high as B−. Roads, water, aviation, transit, and levees all received a D or D− (ASCE 2013).

In 2014 the World Economic Forum (WEF), in its report on global competitiveness, scored countries around the world on the quality of their transportation, electrification, and telephony (Schwab and Sala-i-Martín 2014). The infrastructure factor on which the United States scored highest was the number of available airline seats, for which it was ranked best in the world (although its overall air transport infrastructure was ranked 18th). On the other hand, it was 30th in quality of electricity supply, and 18th in both landline telephone communications and quality of roads. The picture is substantially worse for modern communications infrastructure in the United States: WEF ranks the country 95th in mobile communications and 35th in Internet bandwidth, behind Australia, Barbados, Hong Kong, and much of Western and Eastern Europe.

Improvements in Traditional Infrastructure

One of the best ways that the United States can encourage the creation of value is to upgrade key aspects of its infrastructure. Limitations to the US transportation infrastructure, for example, hurt productivity and result in large costs to the economy. The economy lost an estimated $22 billion from airport congestion and delays in 2012, $90 billion from deficient transit systems, and $101 billion of wasted time and fuel from traffic congestion (ASCE 2013). The

---

[19] Comments by Chad Syverson at meeting of the Foundational Committee on Best Practices for the Making Value for America Study, September 23, 2013, the National Academies, Washington, DC.

nation's port systems, which are critical for the transportation of goods, are also in need of improvement.

Another area that is attracting attention is improvement of the reliability and efficiency of the electricity system. There is work being done, for instance, on the development of smart microgrids; these are much smaller versions of the current centralized systems for generating and transmitting electricity, and they are "smart" in the sense that electrical supply and demand are constantly monitored and regulated to maximize efficiency (Berkeley Lab 2014). To the extent that these microgrids lead to a supply of electricity that is more reliable, more efficient, and greener than the traditional electrical supply, their development could be a competitive advantage for the United States.

Improvements in Information, Communications, and Computing Infrastructure

Generally speaking, any infrastructure improvements that increase the ability of people to communicate and interact are likely to improve the nation's ability to create value.

The committee concluded that one of the most important infrastructure improvements that would enable future value creation in the United States is access to high-speed Internet—particularly wireless—and high-performance computing. As described in Chapter 2, advances in computing power are driving improvements across manufacturing value chains: computer modeling and simulation capabilities increase production efficiency, reduce the need for expensive physical prototyping and testing, and increase quality and reliability along the value chain. Advanced computing capabilities are also enabling entirely new types of products and services. Many of the emerging technologies and capabilities described in Chapter 2—such as data collection, social media analysis, and autonomous vehicles—depend on sophisticated computing capabilities.

Companies and communities are starting to invest in measures to enhance access to high-speed computing capacity. Google operates ultra-high-speed "fiber" services in three US cities—Kansas City, Provo, and Austin—and is considering building such networks in nine more cities (Finley 2014). These services carry data at a gigabit per second, or about 100 times faster than today's typical Internet connections, through a fiber-optic connection directly to customers' homes. Other US municipalities have also installed or are planning to install such ultra-high-speed networks, either themselves or by enlisting a company to build them (Kopytoff 2013). And Google is working on technology that will make it possible to send data across a network at 10 gigabits per second, 1,000 times faster than the typical Internet connection today (Wilke 2014).

One of the most important changes needed to improve wireless is the modernization of spectrum allocation. Use of mobile devices that rely on wireless data and calls has been growing rapidly, but the capacity for these transmissions has not increased because there is little available spectrum to carry them (Rosston 2013). There is, however, significant opportunity to more efficiently allocate spectrum. Spectrum allocation has historically been assigned in an ad hoc manner and could be improved by repurposing the pool of spectrum to make more capacity available for use by high-demand applications such as mobile broadband (Bennett 2012).

Access to reliable, high-speed networks and high-performance computing is essential to improve connectivity and ensure reliable production and service, cornerstones of innovation and value creation.

FEDERAL PROGRAMS THAT MONITOR THE VALUE CHAIN

A variety of federal agencies and programs track the performance of various activities in US-based manufacturing and high-tech service value chains. Two of the most prominent statistical agencies are the Department of Labor (DOL) and Department of Commerce (DOC). The DOL's Bureau of Labor Statistics collects and publicizes labor market information such as employment, pay and benefits, and labor productivity. The DOC's Bureau of Economic Analysis publicizes economic accounts statistics such as gross domestic product, input-output tables, and trade in goods and services. Also housed in DOC, the Census Bureau collects additional information on trade, employment, wages, and a variety of industrial operations, including best management practices. Each of these statistical agencies conforms to the North American Industry Classification System (NAICS), which divides the economy into manufacturing, transportation and warehousing, wholesale trade, retail, professional and business services, information, and ten other industry sectors.

NAICS is particularly important because many datasets, such as the census and economic accounts, rely on this classification system to support government policymaking and inform the American public about the condition of US industries and the overall economy. These datasets are the lens through which policymakers and economists view industrial activity and therefore have a profound influence on the government's and public's understanding of the economy (Dalziel 2007). Statistics based on NAICS are used to monitor the economic status of the United States, determine businesses' eligibility for particular tax exemptions and government contracts, and determine which businesses are subject to certain regulations.

NAICS organizes economic activity based on how establishments carry out their activities, rather than the purpose for those activities, and thus has the advantage of grouping activities that have similar production processes. But it ignores relationships among activities along the same value chain, which traverses the production of goods, services, and software. For example, the system groups automotive and pharmaceutical manufacturers together and the production of automotive electronics with electronic medical equipment, while ignoring the relationship between automotive manufacturers, vehicle electronic component suppliers, dealers, and automotive repair shops.

The fact that the industry classification systems do not account for value chains is increasingly a problem as the use of software, electronic components, and services is becoming more important across a variety of industries, as discussed in chapter 2. The types of goods and services required to meet a particular demand, and how they are produced, have changed enormously since NAICS was established in the 1990s, but the convention of organizing economic metrics by means of production is not flexible to these changes. As a result, a large portion of US economic activity is accounted for by "unmeasurable sectors" (such as the app economy), which are not monitored (Mandel 2012).

Arranging economic statistics instead by the systems of activities along value chains would allow a representation of the economy that reflects the ways companies organize themselves into clusters and sectors (Dalziel 2007). Moreover, it would be less vulnerable to changes in technology than the current approach. Such a classification would also facilitate an understanding of how regulations, economic forces, and other stimuli propagate through interrelated segments of the economy.

REFERENCES

Acemoglu D, Akeigit U, Bloom N, Kerr WR. 2013. Innovation, Reallocation and Growth. NBER Working Paper 18993. Cambridge, MA: National Bureau of Economic Research.

ACSFA [Advisory Committee on Student Financial Assistance]. 2008. Transition matters: Community college to bachelor's degree. Washington. Available at www.ed.gov/acsfa (accessed February 2, 2015).

Aggarwal I, Woolley AW, Chabris CF, Malone TW. 2013. Learning how to coordinate: The moderating role of cognitive diversity on the relationship between collective intelligence and team learning. Carnegie Mellon University Working Paper.

Allen J, Reichheld FF, Hamilton B, Markey R. 2005. Closing the delivery gap: How to achieve true customer-led growth. Bain Insights, October 5. Available at www.bain.com/publications/articles/closing-the-delivery-gap-newsletter.aspx (accessed May 3, 2014).

ASCE [American Society of Civil Engineers]. 2013. 2013 Report Card for America's Infrastructure. Available at www.infrastructurereportcard.org/ (accessed February 2, 2015).

Bakia M, Shear L, Toyama Y, Lasseter A. 2012. Understanding the Implications of Online Learning for Educational Productivity. Washington: Office of Educational Technology, US Department of Education.

Banaji MR, Greenwald AG. 2013. Blindspot: Hidden Biases of Good People. New York: Delacorte Press.

Barta T, Kleiner M, Neumann T. 2012. Is there a payment from top team diversity? McKinsey Quarterly (April): 1–3.

Bartel AP. 1994. Productivity gains from the implementation of employee training programs. Industrial Relations 33:411–425.

Bartel A. 2000. Measuring the employer's return on investments in training: Evidence from the literature. Industrial Relations: A Journal of Economy and Society 39(3):502–524.

Bates TW, Kahle KM, Stulz RM. 2009. Why do US firms hold so much more cash than they used to? Journal of Finance 64(5):1985–2021.

Battelle. 2012. State Bioscience Industry Development 2012. Washington: Battelle Biotechnology Industry Organization.

Becheikh N, Landry R, Amara N. 2006. Lessons from innovation empirical studies in the manufacturing sector: A systematic review of the literature from 1993–2003. Technovation 26(5-6):644–664.

Bennett R. 2012. Powering the Mobile Revolution: Principles of Spectrum Allocation. Washington: Information Technology and Innovation Foundation.

Berkeley Lab. 2014. Microgrids at Berkeley Lab. Available at http://building-microgrid.lbl.gov/ (accessed May 8, 2014).

BEST [Building Engineering & Science Talent]. 2004. A Bridge for All: Higher Education Design Principles to Broaden Participation in Science, Technology, Engineering and Mathematics. San Diego: Council on Competitiveness.

Bhojraj S, Libby R. 2005. Capital market pressure, disclosure frequency-induced earnings/cash flow conflict, and managerial myopia. Accounting Review 80(1):1–20.

Bloom N, Van Reenen J. 2010. Why do management practices differ across firms and countries? Journal of Economic Perspectives 24(1):203–224.

Bloom N, Homkes R, Sadun R, Van Reenen J. 2011. Why American management rules the world. Harvard Business Review, June 13. Available at http://blogs.hbr.org/2011/06/why-american-management-rules/ (accessed May 3, 2014).

Bloom N, Brynjolfsson E, Foster L, Jarmin R, Saporta-Eksten I, Van Reenen J. 2013. Management in America. CES 13-01. Washington: US Census Bureau, Center for Economic Studies. Available at www2.census.gov/ces/wp/2013/CES-WP-13-01.pdf (accessed April 7, 2014).

Brash R, Gallagher M. 2008. A Performance Analysis of SBA's Loan and Investment Programs. Final Report. Washington: Urban Institute.

Brown K. 2014. Shift: More women in computer science classes. San Francisco Chronicle, February 18.

Brown T. 2008. Design Thinking. Harvard Business Review (June):84–92.

Brown T. 2009. Change by Design: How Design Thinking Transforms Organizations and Inspires Innovation. New York: Harper Business.

Butterfield B. 1991. While much of corporate America retreats from affirmative action… Xerox makes it work. Boston Globe, October 20.

Carfang T. 2014. Treasury Strategies: Record High US Corporate Cash Levels Break the $2 Trillion Barrier. New York: Treasury Strategies.

Carlson DS, Upton N, Seaman S. 2006. The impact of human resources practices and compensation design on performance: An analysis of family-owned SMEs. Journal of Small Business Management 44(4):531–543.

Carnevale AP, Rose SJ, Hanson AR. 2012. Certificates: Gateway to gainful employment and college degrees. Center on Education and the Workforce Report, Georgetown University. Available at cew.georgetown.edu/certificates (accessed September 4, 2014).

Černe M, Jaklič M, Škerlavaj M. 2013. Decoupling management and technological innovations: Resolving the individualism-collectivism controversy. Journal of International Management 19(2):103–117.

Chellaraj G, Maskus KE, Mattoo A. 2005. The Contribution of Skilled Immigration and International Graduate Students to US Innovation. Policy Research Working Paper 3588. Washington: World Bank.

Christensen CM, Raynor ME. 2003. The Innovator's Solution: Creating and Sustaining Successful Growth. Cambridge, MA: Harvard Business School Press.

Cowen T. 2011. The Great Stagnation: How America Ate All the Low-Hanging Fruit of Modern History, Got Sick, and Will (Eventually) Feel Better. New York: Dutton.

Dalziel M. 2007. A systems-based approach to industry classification. Research Policy 36:1559–1574.

David E, Mehta A, Norris T, Singh N, Tramontin T. 2010. New frontiers in pharma R&D investment. McKinsey Quarterly (February):1–12.

Davis SJ, Haltiwanger J. 2014. Labor Market Fluidity and Economic Performance. NBER Working Paper No. 20479. Cambridge, MA: National Bureau of Economic Research.

Decker R, Haltiwanger J, Jarmin R, Miranda J. 2014. The role of entrepreneurship in US job creation and economic dynamism. Journal of Economic Perspectives 28(3):3–24.

Decker WH, Calo TJ, Weer CH. 2012. Affiliation motivation and interest in entrepreneurial careers. Journal of Managerial Psychology 27(3):302–320.

Dossani R. 2014. What to make of P-TECH schools. The RAND Blog, February 21. Available at www.rand.org/blog/2014/02/what-to-make-of-p-tech-schools.html (accessed May 6, 2014).

Dutta D, Patil L, Porter JB Jr. 2012. Lifelong Learning Imperative in Engineering: Sustaining American Competitiveness. Washington: National Academies Press.

Earle JS, Sakova Z. 2000. Business start-ups or disguised unemployment? Evidence on the character of self-employment from transition economies. Labour Economics 7:575–601.

Economist. 2012. All together now: The advantages of crowdsourcing. Special report, April 21. Available at www.economist.com/node/21552902 (accessed May 3, 2014).

EOP [Executive Office of the President]. 2014. Report to the President: Accelerating US Advanced Manufacturing. Washington.

Fairlie RW. 2014. Kauffman Index of Entrepreneurial Activity 1996–2013. Kansas City, MO: Ewing Marion Kauffman Foundation.

Finley K. 2014. Google explores super-speed Internet in 9 more cities. Wired, February 19. Available at www.wired.com/2014/02/google-fiber-cities/ (accessed May 8, 2014).

Forbes. 2011. Global diversity and inclusion: Fostering innovation through a diverse workforce. Available at http://images.forbes.com/forbesinsights/StudyPDFs/Innovation_Through_Diversity.pdf (accessed February 2, 2014).

Ford D, Nelsen B. 2013. The view beyond venture capital. Nature Biotechnology 32:15–23. Available at www.nature.com/nbt/journal/v32/n1/full/nbt.2780.html (accessed February 2, 2015).

Foroohar R. 2014. The school that will get you a job. Time, February 13. Available at http://time.com/7066/the-school-that-will-get-you-a-job/ (accessed May 6, 2014).

Ganzglass E. 2014. Scaling "Stackable Credentials": Implications for Implementation and Policy. Washington: Center for Law and Social Policy. Available at www.clasp.org (accessed February 2, 2015).

Ganzglass E, Bird K, Prince H. 2011. Giving Credit Where Credit Is Due: Creating a Competency-Based Qualifications Framework for Postsecondary Education and Training. Washington: Center for Law and Social Policy. Available at www.clasp.org (accessed February 4, 2015).

Gorodnichenko Y, Roland G. 2011. Individualism, innovation, and long-run growth. Proceedings of the National Academy of Sciences 108(Supplement 4):21316–21319.

Greenstone M, Looney A. 2011. Building America's Job Skills with Effective Workforce Programs: A Training Strategy to Raise Wages and Increase Work Opportunities. Strategy Paper, The Hamilton Project. Washington: Brookings Institution.

Haltiwanger J, Jarmin R, Miranda J. 2012. Where Have All the Young Firms Gone? Business Dynamics Statistics Briefing 6. Washington: US Bureau of the Census.

Haltiwanger J, Jarmin RS, Miranda J. 2013. Who creates jobs? Small versus large versus young. Review of Economics and Statistics 95(2):347–361.

Hansson B. 2007. Company-based determinants of training and the impact of training on company performance: Results from an international HRM survey. Personnel Review 36(2):311–331.

Harrington B, Ladge JJ. 2009. Got talent? It isn't hard to find: Recognizing and rewarding the value women create in the workplace. The Shriver Report, pp. 199–131. Available at http://shriverreport.org/got-talent-it-isnt-hard-to-find/ (accessed February 6, 2015).

Haskins R, Holzer H, Lerman R. 2009. Promoting Economic Mobility by Increasing Secondary Education. Philadelphia: Pew Charitable Trusts.

Herring C. 2009. Does diversity pay? Race, gender, and the business case for diversity. American Sociological Review 74(2):208–224.

Holzer HJ, Dunlop E. 2013. Just the Facts, Ma'am: Postsecondary Education and Labor Markets in the US. IZA Discussion Paper No. 7319. Bonn: Institute for the Study of Labor. Available at http://papers.ssrn.com/sol3/papers.cfm?abstract_id=2250297 (accessed February 4, 2015).

Hong L, Page SE. 2004. Groups of diverse problem solvers can outperform groups of high-ability problem solvers. Proceedings of the National Academy of Sciences 101(46):16385–16389.

Hoogendoorn S, Oosterbeek H, Van Praag M. 2013. The impact of gender diversity on the performance of business teams: Evidence from a field experiment. Management Science 59(7):1514–1528.

Huggins R. 2001. Inter-firm network policies and firm performance: Evaluating the impact of initiatives in the United Kingdom. Research Policy 30:443–458.

Hunt J, Gauthier-Loiselle M. 2010. How much does immigration boost innovation? American Economic Journal: Macroeconomics 2(2):31–56.

ICF. 2009. Corridor Approaches to Integrating Transportation and Land Use. Washington.

Ichniowski C, Shaw K, Prennushi G. 1995. The Effects of Human Resource Management Practices on Productivity. NBER Working Paper 5333. Cambridge, MA: National Bureau of Economic Research.

Inkpen AC. 2005. Learning through alliances: General Motors and NUMMI. California Management Review 47(4):114.

Johansson F. 2004. The Medici Effect: What Elephants and Epidemics Can Teach Us about Innovation. Cambridge, MA: Harvard Business School Press.

Kahneman D. 2011. Thinking, Fast and Slow. New York: Farrar, Straus and Giroux.

Katz B, Wagner J. 2014. The Rise of Innovation Districts: A New Geography of Innovation in America. Washington: Brookings Institution.

Keeley L, Walters H, Pikkel R, Quinn B. 2013. Ten Types of Innovation: The Discipline of Building Breakthroughs. New York: Wiley.

Kelley T. 2001. The Art of Innovation: Lessons in Creativity from IDEO, America's Leading Design Firm. New York: Crown Business.

Khanna D. 2013. We need more women in tech: The data prove it. The Atlantic, October 29.

Kingsley G, Klein HK. 1998. Interfirm collaboration as a modernization strategy: A survey of case studies. Journal of Technology Transfer 23(1):65–74.

Klobuchar A. 2013. Women in Manufacturing. Washington: Joint Economic Committee, US Senate.

Kopytoff V. 2013. Google's not the only one with super-high-speed Internet plans. Fortune, June 18. Available at http://tech.fortune.cnn.com/2013/06/18/googles-not-the-only-one-with-super-high-speed-internet-plans/ (accessed May 8, 2014).

Krueger A, Rouse C. 1998. The effect of workplace education on earnings, turnover, and job performance. Journal of Labor Economics 16(1):61–94.

Leachman M, Mai C. 2014. Most States Funding Schools Less than Before the Recession. Washington: Center on Budget and Policy Priorities. Available at www.cbpp.org/files/9-12-13sfp.pdf (accessed February 2, 2015).

Locke RM, Wellhausen RL. 2014. Production in the Innovation Economy. Cambridge, MA: MIT Press.

Lovett M, Meyer O, Thille C. 2008. The open learning initiative: Measuring the effectiveness of the OLI statistics course in accelerating student learning. Journal of Interactive Media in Education (1):13.

Lund S, Manyika J, Nyquist S. 2013a. Breaking the US growth impasse. McKinsey Quarterly (August):1–7.

Lund S, Manyika J, Nyquist S, Mendonca L, Ramaswamy S. 2013b. Game changers: Five opportunities for US growth and renewal. Washington: McKinsey Global Institute.

Lynch LM. 2004. Development intermediaries and the training of low-wage workers. In: Emerging Labor Market Institutions for the Twenty-First Century, ed. RB Freeman, J Hersch, L Mishel. Cambridge, MA: National Bureau of Economic Research. Available at www.nber.org/chapters/c9959 (accessed February 4, 2015).

Lynn BC, Khan L. 2012. The slow-motion collapse of American entrepreneurship. Washington Monthly (July/August). Available at www.washingtonmonthly.com/magazine/july august_2012/features/the_slowmotion_collapse_of_ame038414.php?page=all (accessed February 2, 2015).

MacLeod G. 1997. From Mondragon to America: Experiments in Community Economic Development. Sydney, Nova Scotia: University College of Cape Breton Press.

Mandel M. 2012. Where the Jobs Are: The App Economy. Washington: TechNet.

Manyika J, Sinclair J, Dobbs R, Strube G, Rassey L, Mischke J, Remes J, Roxburgh C, George K, O'Halloran D, Ramaswamy S. 2012. Manufacturing the Future: The Next Era of Global Growth and Innovation. London: McKinsey Global Institute. Available at www.mckinsey.com/insights/manufacturing/the_future_of_manufacturing (accessed April 4, 2014).

Margolis RM, Kammen DM. 1999. Evidence of under-investment in energy R&D in the United States and the impact of federal policy. Energy Policy 27(10):575–584.

Martin R. 2009. The Design of Business: Why Design Thinking Is the Next Competitive Advantage. Cambridge, MA: Harvard Business School Press.

Mitchell L. 2011. Overcoming the Gender Gap: Women Entrepreneurs as Economic Drivers. St. Louis, MO: Ewing Marion Kauffman Foundation.

Moran N. 2007. Technology commercialization firms float in UK—but not elsewhere. Nature Biotechnology 25(7):697–698.

NAE [National Academy of Engineering]. 2005. Engineering Research and America's Future: Meeting the Challenges of a Global Economy. Washington: National Academies Press.

Narayanamurti V, Odumosu T, Vinsel L. 2013. RIP: The basic/applied research dichotomy. Issues in Science and Technology 29(2).

Niebuhr A. 2010. Migration and innovation: Does cultural diversity matter for regional R&D activity? Papers in Regional Science 89(3):563–585.

NRC [National Research Council]. 1994. Women Scientists and Engineers Employed in Industry: Why So Few? Washington: National Academy Press.

NRC. 2002. Government-Industry Partnerships for the Development of New Technologies. Washington: National Academies Press.

NRC. 2008. Innovative Flanders: Innovation Policies for the 21st Century: Report of a Symposium. Washington: National Academies Press.

NRC. 2009. Venture Funding and the NIH SBIR Program. Washington: National Academies Press.

NRC. 2011. Successful K-12 STEM Education: Identifying Effective Approaches in Science, Technology, Engineering, and Mathematics. Washington: National Academies Press.

NRC. 2012a. Research Universities and the Future of America: Ten Breakthrough Actions Vital to Our Nation's Prosperity and Security. Washington: National Academies Press.

NRC. 2012b. Improving Measurement of Productivity in Higher Education. Washington: National Academies Press.

NRC. 2013a. Next Generation Science Standards: For States, By States. Washington: National Academies Press.

NRC. 2013b. New York's Nanotechnology Model: Building the Innovation Economy: Summary of a Symposium. Washington: National Academies Press.

NRC. 2013c. Strengthening American Manufacturing: The Role of the Manufacturing Extension Partnership: Summary of a Symposium. Washington: National Academies Press.

NSB [National Science Board]. 2012. Science and Engineering Indicators 2012. NSB 12-01. Arlington, VA: National Science Foundation.

NVCA [National Venture Capital Association]. 2013. Patient Capital 3.0. NVCA/MedIC Coalition. Washington.

OECD [Organization for Economic Cooperation and Development]. 2009. Top barriers and drivers to SME internationalization. Report by the OECD Working Party on SMEs and Entrepreneurship. Paris. Available at www.oecd.org/cfe/smes/43357832.pdf (accessed July 15, 2014).

Ottaviano GI, Peri G. 2006. The economic value of cultural diversity: Evidence from US cities. Journal of Economic Geography 6(1):9–44.

Panagiotakopoulos A. 2011. Barriers to employee training and learning in small and medium-sized enterprises (SMEs). Development and Learning in Organizations 25(3):15–18.

Parrotta P, Pozzoli D, Pytlikova M. 2012. The nexus between labor diversity and firm's innovation. Discussion Paper Series, No. 6972. Bonn: Institute for the Study of Labor. Available at http://hdl.handle.net/10419/67257 (accessed February 5, 2015).

Pine J, Tingley JC. 1993. ROI of soft-skills training. Training 30(2):55–58.

PNAE [Partnership for a New American Economy]. 2012. How Immigrants Are Reinventing the American Economy. New York. Available at www.renewoureconomy.org/sites/all/themes/pnae/patent-pending.pdf (accessed October 21, 2014).

Ramamoorthy N, Flood PC, Slattery T, Sardessai R. 2005. Determinants of innovative work behaviour: Development and test of an integrated model. Creativity and Innovation Management 14(2):142–150.

Ries E. 2011. The Lean Startup: How Today's Entrepreneurs Use Continuous Innovation to Create Radically Successful Businesses. New York: Crown Business.

Rosston GL. 2013. Increasing Wireless Value: Technology, Spectrum, and Incentives. Stanford Institute for Economic Policy Research.

Schwab K, Sala-i-Martín X, eds. 2014. The Global Competitiveness Report, 2013–2014: Full Data Edition. Geneva: World Economic Forum.

Shane S. 1993. Cultural influences on national rates of innovation. Journal of Business Venturing 8(1):59–73.

Steele CM. 2010. Whistling Vivaldi: And Other Clues to How Stereotypes Affect Us. New York: WW Norton & Company.

Stein JC. 1989. Efficient capital markets, inefficient firms: A model of myopic corporate behavior. Quarterly Journal of Economics 104(4):655–699.

Syverson C. 2011. What determines productivity? Journal of Economic Literature 49(2):326–365.

Taylor MZ, Wilson S. 2012. Does culture still matter? The effects of individualism on national innovation rates. Journal of Business Venturing 27(2):234–247.

Tiessen JH. 1997. Individualism, collectivism, and entrepreneurship: A framework for international comparative research. Journal of Business Venturing 12(5):367–384.

Tsapogas J. 2004. The Role of Community Colleges in the Education of Recent Science and Engineering Graduates. NSF 04-315. Arlington, VA: National Science Foundation.

Veum JR. 1995. Sources of training and their impact on wages. Industrial and Labor Relations Review 48(4):812–826.

Wadhwa V, Saxenian A, Rissing B, Gereffi G. 2007. America's New Immigrant Entrepreneurs. Duke Science, Technology, & Innovation Paper No. 23. Durham: Duke University School of Law. Available at http://ssrn.com/abstract=990152 (accessed February 6, 2015).

Webb M, Gerwin C. 2014. Early College Expansion: Propelling Students to Postsecondary Success, at a School Near You. Washington: Jobs for the Future. Available at www.jff.org/publications/early-college-expansion-propelling-students-postsecondary-success-school-near-you (accessed February 4, 2015).

WEF [World Economic Forum]. 2013. Global Competitiveness Report 2013–2014. New York.

Welch D, Oldsman E, Shapira P, Youtie J, Lee J. 1997. Net Benefits: An Assessment of a Set of Manufacturing Business Networks and Their Impacts on Member Companies. USNet Evaluation Working Paper 9701. Atlanta: Georgia Institute of Technology.

Wilke N. 2014. Google developing Internet that's over 1,000 times faster than yours. Wired, February 14. Available at www.wired.com/2014/02/100-gigabits/ (accessed May 8, 2014).

Womack JP, Jones DT, Roos D. 1990. The Machine That Changed the World: Based on the Massachusetts Institute of Technology 5-Million-Dollar 5-Year Study on the Future of the Automobile. Riverside, NJ: Simon & Schuster.

Woolley AW, Malone TW. 2011. What makes a team smarter? More women. Harvard Business Review, June. Available at http://hbr.org/2011/06/defend-your-research-what-makes-a-team-smarter-more-women/ar/1 (accessed February 3, 2014).

Woolley AW, Chabris CF, Pentland A, Hashmi N, Malone TW. 2010. Evidence for a collective intelligence factor in the performance of human groups. Science 330(6004):686–688.

Xerox. 2013. 2013 Summary Report on Global Citizenship. Norwalk, CT.

4

Creating a Prosperous Path Forward: Recommended Actions

To prosper in the 21st century, US companies and communities must take action to strengthen the country's capacity for innovation along the manufacturing value chain. Businesses have opportunities to take individual action to improve their competitiveness. A variety of stakeholders also have important roles to play to ensure that the United States has a robust innovation ecosystem to support manufacturing value chains and the broader economy; these stakeholders include federal, state, and local governments; economic development organizations; educational institutions; and research organizations. This chapter presents the committee's recommendations of specific actions for the various stakeholders in each of the fundamental areas described in Chapter 3 to ensure US leadership in making value along the manufacturing value chain.

ACTIONS TO FACILITATE THE ADOPTION OF BUSINESS BEST PRACTICES

Individual businesses can create value by coordinating their value chains and optimizing their operations.

Businesses across the value chain need to reengineer their operations and adopt best practices to improve innovation, productivity, and speed to market. While every business aims to optimize its operations for productivity, very few have implemented the advanced practices necessary to achieve world-leading productive operations. Businesses also need to leverage technology and talent to ensure a sustainable stream of new products and services, better understand customer needs, and identify value creation opportunities. To accomplish these aims, companies should actively encourage their employees to continually improve operations, identify new market opportunities, and implement the resources needed to commercialize solutions.

Recommendations

- **Businesses should establish training programs to prepare workers for modernized operations and invest in advancing the education of their low- and middle-skilled workforce.** Employee training programs raise earnings potential and substantially increase productivity, profits, and innovation across businesses. Employers gain large returns from investing in the development of their low- and middle-skilled workforce, in some cases as high as 100–200 percent.
- **Companies should examine their business models to search for missed opportunities to leverage distributed tools and coordinate manufacturing and product lifecycle services.** Radical gains come from producing new solutions not provided by others. The ability to provide such solutions requires understanding customer needs and desires and developing an innovation strategy that differentiates a business's offerings from those of its competitors. Organizations will attain a competitive advantage if they understand how economic forces are shifting and can both coordinate capabilities across value chains and leverage digital and distributed tools to generate innovative solutions.
- **Manufacturers should implement principles and practices such as Lean Production that enable employees to improve productivity and achieve continuous improvement.** Systems of best practices for production, such as lean manufacturing, have been shown to improve productivity, decrease time from customer order to delivery, and reduce costs. Best practices such as lean manufacturing can directly benefit a manufacturer's bottom line and create a significant competitive advantage. They can also reduce energy and resource consumption and in some cases make domestic production for the US market more cost-effective than producing abroad.
- **Researchers should further investigate and codify best practices for innovation and develop effective methods of teaching them.** Additional research is needed to identify best practices for identifying unmet needs and commercializing solutions that apply to a wide range of companies and industries. Creating teachable systems of best practices and encouraging their widespread adoption will be important to value creation.

ACTIONS TO ENSURE AMERICA HAS AN
INNOVATIVE WORKFORCE

**The education and skills of the US workforce must be improved.
Higher education and training are increasingly important
to create an effective ecosystem for value creation.**

Maximizing the ability of the United States to create value requires maximizing the development of its talent. The following actions can promote the education needed to prepare US students and workers to compete effectively in innovation and value creation.

Recommendations

- **Businesses, local school districts, labor, community colleges, and universities should form partnerships to help students graduate from high school, earn an associate's or bachelor's degree, and take part in continuing education in the workplace.** More needs to be done to advance the skills of lower-income Americans, and these partnerships can provide students from these families with the guidance, social supports, and education necessary to complete a college degree. State governments should facilitate connections between school districts, businesses, and other actors to form these partnerships.

  There are a number of successful public-private partnerships that can serve as examples to those interested in establishing others. The model that the committee believes is particularly promising is a collaboration of employers and school districts combining classroom-based learning with work experiences to help students graduate from high school and put them on track to a college degree and a skilled job.

- **Congress and state legislatures should create incentives for businesses to invest and be involved in education programs.** Congress and state governments should create tax credits or other incentives to encourage investments in educational partnerships involving businesses, community colleges, and universities to provide students and displaced workers with the knowledge and skills needed for higher-paying careers.

- **Middle schools, high schools, universities, and local communities should provide opportunities for students to participate in team-based engineering design experiences and learn to use emerging digital and distributed tools.** Students exposed to such team-based experiences are better prepared to contribute to today's innovation workforce. Efforts to teach team-based engineering design skills, such as in the framework provided by the Next Generation Science Standards, should be widely adopted.

Extracurricular programs that allow students to participate in team-based design experiences and use emerging tools that enable new business creation should also be expanded.

- **Universities and community colleges should improve the cost-effectiveness of higher education.** Reducing financial barriers for lower-income students will increase the level of talent across the workforce and thus stimulate value creation. Universities and colleges should facilitate students' transfer from two-year community college programs to reduce the costs of a four-year degree. They should also seek opportunities to adopt new methods of teaching—such as online tutorials, computer-based instant feedback on homework assignments, open-access course materials for instructors, and credit-by-examination approaches (which allow students to test out of courses)—to support learning while reducing students' costs.

- **University rating organizations should track and make transparent the cost-effectiveness of degrees at higher education institutions.** Students should have access to information comparing the cost-effectiveness of particular degrees at different universities and colleges along such metrics as the debt and earnings of graduates, attrition rates, and average time to degree. Colleges and universities should be encouraged to use these data to assess and, as appropriate, improve the cost-effectiveness of their degrees.

- **Businesses, industry associations, and higher education institutions should work together to (1) establish national skills certifications that are widely recognized by employers and count toward degree programs, and (2) improve access for students and workers to acquire these certifications.** Standard skills certifications will allow employers to identify the skills of job candidates using a consistent baseline. Credentials are also important to students who may not have completed all the requirements for a college degree but have acquired skills along the way. "Stackable" credentials (those that can be accumulated serially over time) provide an alternative qualification for such students to apply for higher-paying jobs; the Manufacturing Institute has developed a Manufacturing Skills Standard Certification System that recognizes stackable credentials. Manufacturers should recognize these credentials and education institutions should work with manufacturers to both provide access for students and workers to obtain these credentials and recognize them as counting toward a formal degree.

    National skills certifications can and should be established in other areas along the value chain, such as for software specialists. Businesses, industry associations, and education institutions should work together to define stackable credentials and skills certifications in new areas and establish methods of evaluating competencies.

In today's globalized economy, US companies need the best teams in the world to stay competitive. Such teams will depend on not only creating and attracting top talent but also leveraging diversity to achieve better team performance.

The nation's ability to create value will be enhanced by innovative teams that create new and improved products, services, and processes and bring them to market. The inclusion of diverse individuals and the attraction and retention of talent from around the world are critical to ensure that the best teams assemble in the United States. Some groups—particularly women, racial minorities, and people from low-income families—remain significantly less able than others to take advantage of value creation opportunities, whether because of unequal educational opportunities, the lingering effects of historical inequity, or discrimination. This inequity negatively impacts the nation's prosperity, not only because fewer people can become the innovators that create economic growth and jobs but also because teams of people that have more women and people with diverse characteristics have been shown to be more innovative.

In addition to leveraging the strengths of its diverse population, America can improve its ability to create value by attracting and retaining talented people from other countries. The United States already attracts students from all over the world for postgraduate education in the STEM disciplines, but because of current US immigration policies not all of the graduates who wish to stay and work are allowed that option. Making it easier for them to stay could greatly increase the number of people who can contribute to value creation in this country, thus improving the economy for all.

Recommendations

- **Businesses should implement programs to attract and retain diverse workers, including along gender, race, and socioeconomic background.** Increasing this diversity is not only ethical; it is good for innovation and business success. Businesses should implement programs to improve equitable access in hiring and promotions. Efforts to encourage managers to support the career development of their employees and to ensure transparency in internal job opportunities and promotion criteria have significantly improved recruitment and retention of women and underrepresented groups.

- **Universities and community colleges should act to improve the inclusion of traditionally underrepresented groups in science, technology, engineering, and mathematics (STEM) fields as well as other disciplines required for value creation, such as market analysis and design.** Educational institutions should study and learn from approaches that have been successful in attracting these groups to university programs and businesses. STEM programs such as those at the University of California, Berkeley, Carnegie

Mellon University, and Harvey Mudd College, all of which have achieved high enrollment of women and other underrepresented groups, can be used to guide others. All programs should implement known practices that help attract these groups, such as redesigning courses to emphasize the real-world relevance of the material.

- **Congress must reform immigration policy to welcome and retain high-skilled individuals with advanced STEM degrees, especially those educated in the United States.** Many of these individuals become entrepreneurs and the United States should ensure that their businesses are in this country. Unfortunately, however, these potential innovators are being turned away by a counterproductive immigration system. In both 2013 and 2014, the allotment of H-1B visas was filled the first week they were made available.

ACTIONS TO STRENGTHEN LOCAL INNOVATION NETWORKS

The United States needs to encourage new business creation across the value chain to stimulate innovation and job creation.

Business creation across the value chain and the broader economy is critical for the US economy. Statistics indicating that the rate of business creation has been declining in the United States for the past three decades are worrisome.

Recommendation

- **Researchers, the National Science Foundation, and other research funders should put a priority on understanding the declining rate of new business creation.** Data suggest that the rate of new business creation is declining. If so, researchers need to investigate the causes. They should examine whether barriers have increased for new business creation or whether the current environment favors established businesses. Researchers should also investigate the factors that encourage the formation of new businesses and increase their likelihood of success.

Local innovation networks are needed across the United States to foster the creation of new businesses and connect entrepreneurs and new businesses to the individuals, investors, tools, and institutions in their region and around the world that they need to grow.

For the greatest chance of success, potential innovators need to be able to connect with a wide variety of people and organizations, including other innovators, scientists and engineers, investors, workers with useful skills, organizational and management advisors, market analysts and marketing specialists, policymakers, and potential customers. They also need access to the tools, such as prototyping or testing equipment, that enable value creation. The places most supportive of innovators—and thus most likely to see innovation-driven growth—are those that have well-developed networks of investors, academia, industry, sources of financing, sources of business advice, access to required tools, government agencies, nongovernmental organizations, and customers.

Recommendations

- **Metro area and state governments, industry, higher education, investors, and economic development organizations should partner to create local innovation networks.** Any one of these stakeholders can spearhead the creation of such a network. Innovation networks in Silicon Valley, the Boston-Cambridge area, the San Francisco Bay area, Seattle, New York state, Israel, and Singapore can inform efforts to create others. In addition to providing resources for innovators, these networks should support them by, for example, facilitating the sharing of best practices and helping small businesses learn how to export.
- **Metro area and state governments should optimize their decision-making process for urban development investments and siting to facilitate the creation of innovation networks.** In most metro areas, decisions on urban development investment and siting are the responsibility of individual units with different functional missions (e.g., housing, transportation) without the coordinating oversight of a single body. These units need to coordinate their decisions to nurture innovation networks. Cities, surrounding counties, and states should identify opportunities to better structure these decisions to serve the welfare of the entire area.

US programs that contribute to innovation should be directed and optimized as appropriate to facilitate the adoption of best practices and help young businesses to grow.

Recommendations

- **Federal agencies and interagency offices such as the Advanced Manufacturing National Program Office should convene stakeholders to identify and spread best practices for value creation.** AMNPO and other federal agencies should use their convening power to support collaborations that can help identify and spread best practices for 21st century value creation, particularly for software, user interfaces, and high-tech services, where best practices are less developed than production. Companies should be encouraged to collaborate in sharing and sharpening best practices and solving common problems, spreading the cost of finding solutions, and stimulating the movement of ideas across industries.

- **The Small Business Administration should help more young businesses become globally competitive.** The SBA should continue to help businesses become globally competitive, recognizing that young businesses in particular are some of the fastest-growing companies and are potentially the most responsive to influxes of financial capital. In particular, it should help young businesses connect with a local innovation network, and if one does not exist it should encourage the formation of one.

ACTIONS TO FACILITATE THE FLOW OF CAPITAL INVESTMENTS

US fiscal policy must incentivize long-term capital investments.

Increasing emphasis on short-term returns on investment has led to a decrease in the long-term planning and funding necessary to support many promising innovations. New models are needed to ensure the long-term investments necessary to develop groundbreaking innovations.

Recommendations

- **Congress should modify the capital gains tax rates to incentivize holding stocks for five years, ten years, and longer.** The current tax structure encourages a preference for quicker returns over the long-term investments needed to create new products and businesses. Capital gains tax rates do not provide incentives for investments longer than one year. Congress should create favorable tax treatment for stocks held for five years, ten years, and longer. It should also identify and implement additional opportunities in the tax code to enable access to both short- and long-term low-cost capital.
- **Congress should make the research-and-development tax credit permanent to encourage businesses toward longer-term horizons in their investment decisions.** The tendency of Congress to renew the tax credit for only two years discourages businesses from investing in longer-term R&D projects. A permanent tax credit would stimulate R&D spending, thereby increasing economic growth and fostering innovation.
- **Federal agencies should facilitate industry and government cooperation to identify shared opportunities to invest in precompetitive research in long-term, capital-intensive fields** such as next-generation batteries and biotechnologies, for which capital availability is scarce.

ACTIONS TO PROVIDE AN INFRASTRUCTURE
THAT ENABLES VALUE CREATION

US infrastructure must be upgraded for both traditional systems (e.g., electricity, ports) and modern information systems. A world-leading infrastructure will attract businesses and facilitate the creation of new ones in the United States.

Infrastructure is crucial to innovation and value creation. Countries whose infrastructure is deficient (or even lacking) with respect to objective benchmarks with the rest of the world will find it more difficult to innovate and create value. A country that hopes to be at the forefront of innovation must be at the forefront in terms of its infrastructure.

Recommendations

- **Local governments, state legislatures, and Congress should invest in a world-leading wireless infrastructure.** Infrastructure that makes it easier for individuals and machines to communicate and process information is essential for innovation along the value chain. Innovation is almost always a team effort, which requires the seamless ability to exchange ideas and information. Furthermore, many emerging technologies and service improvements rely on real-time information collection, processing, and transmission.

- **Federal information technology and computing programs should facilitate access to a world-leading infrastructure for high-performance computing.** High-performance computing capabilities can drive improvements across the value chain and enable entirely new types of products and services. These resources require substantial investment and are therefore not always accessible, especially to small businesses. Federal agencies should work to improve access to high-performance computing.

ACTIONS TO IMPROVE METHODS OF MONITORING
MANUFACTURING VALUE CHAINS

Federal programs and statistics should be modernized to account for the diminishing distinction and complex relationships between manufacturing, information, and services.

Modern value chains often involve a complex network of activities that span the classic economic sectors. The production of raw materials, goods, services, and software are interconnected along these value chains, and often carried out within the same business or even a single establishment. As businesses traditionally known for manufacturing move into software and service production, and companies known for creating software and online services produce manufactured goods, it is increasingly difficult to meaningfully delineate operations as providing mainly goods or services.

Recommendations

- **Federal agencies should develop methods of accounting for the complex relationships between manufacturing, services, and information and consider multiple ways of collecting and organizing national statistics.** Agencies that collect economic statistics, such as the Bureau of Labor Statistics and the Bureau of Economic Analysis, should identify methods to (1) capture the complex relationships between industry sectors and (2) organize national statistics in a way that complements the current classification. The current method of organizing national economic statistics— classifying manufacturing, services, and information activities in distinct industries based on the primary activity at an establishment—is an increasingly unrealistic depiction. Such a system does not provide any information about services and information activities undertaken by manufacturers or production operations that are primarily carried out to support a software or service provider. It also does not allow for an understanding of value chains, such as the manufacturing, service, and information operations devoted to improving health outcomes or providing personal transportation. This is particularly problematic because these statistics influence policy decisions that affect innovation and education that would benefit from an understanding of these nuances.

- **Federal programs that contribute to innovation should be directed and optimized as appropriate to assist software and service providers as well as manufacturers.** Federal programs to revitalize manufacturing in the United States, such as the Advanced Manufacturing National Program Office (AMNPO), the Manufacturing Extension Partnership, and the Advanced Manufacturing Partnership, should not lose sight of the importance of software and service providers. Software and services are integral to manufacturing value chains and increasingly important to businesses' capacity to take advantage of emerging digital technologies. The administration and federal agencies should review and optimize current programs to ensure that all activities across the value chain are appropriately supported.

Table 4-1 compiles the committee's recommendations directed to businesses, the federal government, state governments, localities, education institutions, and other actors.

TABLE 4-1 Recommendations organized by actor

Actor	Recommendations
Businesses	• Companies should examine their business models to search for missed opportunities to leverage distributed tools and coordinate manufacturing and product lifecycle services. • Businesses should establish training programs to prepare workers for modernized operations and invest in advancing the education of their low- and middle-skilled workforce. • Manufacturers should implement principles and practices such as Lean Manufacturing that enable employees to improve productivity and achieve continuous improvement. • Businesses should work with local school districts, community colleges, and universities to form partnerships to help students graduate from high school, earn an associate's or bachelor's degree, and take part in continuing education in the workplace. • Businesses should work with industry associations and higher education institutions to (1) establish national skills certifications that are widely recognized by employers and count toward degree programs, and (2) improve access for students and workers to gain these certifications. • Businesses should attract and implement programs to retain diverse workers, including along gender, race, and socioeconomic background.
Federal government	• Federal agencies and interagency offices such as the Advanced Manufacturing National Program Office should convene stakeholders to identify and spread best practices for value creation. • Congress should establish incentives for businesses to invest and be involved in education programs. • Congress must reform immigration policy to welcome and retain high-skilled individuals with advanced STEM degrees, especially those educated in the United States. • The Small Business Administration should focus on helping young businesses become globally competitive as opposed to focusing on older, established small businesses. • The National Science Foundation and other research funders should put a priority on research to understand the declining rate of new business creation. • Federal programs that contribute to innovation should be directed and optimized as appropriate to assist software and service providers as well as manufacturers. • Congress should modify the capital gains tax rates to incentivize holding stocks for five years, ten years, and longer.

TABLE 4-1 Continued

Actor	Recommendations
Federal government (continued)	• Congress should make the research-and-development tax credit permanent to encourage businesses to adopt longer-term horizons in their investment decisions.
	• Federal agencies should facilitate industry and government cooperation to identify shared opportunities to invest in precompetitive research in long-term, capital-intensive fields.
	• Congress should support state legislatures and local governments to invest in a world-leading wireless infrastructure.
	• Federal information technology and computing programs should facilitate access to a world-leading infrastructure for high-performance computing.
	• Federal agencies should develop methods of accounting for the complex relationships between manufacturing, services, and information and consider multiple ways of collecting and organizing national statistics.
State governments	• State governments should establish incentives for businesses to invest and be involved in education programs.
	• State governments should partner with local governments, industry, higher education, investors, and economic development organizations to create local innovation networks.
	• State governments should work with local governments to optimize the decision-making process for urban development investments and siting to facilitate the creation of innovation networks.
	• State legislatures, with local government and Congressional support, should invest in a world-leading wireless infrastructure.

continued

TABLE 4-1 Continued

Actor	Recommendations
Local governments	• Local school districts should work with businesses and community colleges to form partnerships to help students graduate from high school, earn an associate's or bachelor's degree, and take part in continuing education in the workplace. • Metro area governments should partner with state governments, industry, higher education, investors, and economic development organizations to create local innovation networks. • Metro area governments should work with state governments to optimize the decision-making process for urban development investments and siting in order to facilitate the creation of innovation networks. • Local governments, with state government and Congressional support, should invest in a world-leading wireless infrastructure.
Education institutions	• Community colleges and universities should partner with local school districts and businesses to help students graduate from high school, earn an associate's or bachelor's degree, and take part in continuing education in the workplace. • Middle schools, high schools, and local communities should provide opportunities for students to participate in team-based engineering design experiences and learn how to use emerging tools that enable new business creation. • Universities and community colleges should improve the cost-effectiveness of higher education. • Higher education institutions should work together with businesses and industry associations to (1) establish national skills certifications that are widely recognized by employers and count toward degree programs, and (2) improve access for students and workers to gain these certifications. • Universities and community colleges should act to improve the inclusion of traditionally underrepresented groups in science, technology, engineering, and mathematics (STEM) fields as well as other disciplines required for value creation, such as market analysis and design.
Other actors	• Researchers should further investigate and codify best practices for innovation and develop effective methods of teaching them. • University rating organizations should track and make transparent the cost-effectiveness of degrees at higher education institutions.

Appendix

The Big Picture

The changes affecting manufacturing value chains underscore the importance of creating an environment in the United States that continuously attracts and creates businesses and jobs. If the nation is to replace jobs that have been disrupted along the value chain, it will need to be in a strong position for its businesses to compete globally. Looking at the current state of activities in US-based value chains in the context of the global economy and the country's ability to attract and create businesses and jobs, three challenges are apparent. First, there is growing competition from countries around the world. Second, there are concerns that the United States is falling behind in some of the critical inputs for value creation, such as capital investments and student learning in science, technology, engineering, and math. Last, the birth rate of new businesses across the value chain—in production, retail, and services—has been declining in the United States.

While the development of economies around the world has intensified competition, it also presents enormous opportunities to expand demand for US goods and services. Emerging markets offer tremendous potential for US companies in the coming decades, but only if companies and policymakers recognize the potential and develop and maintain the capabilities to take advantage of it.

US POSITION IN GLOBAL INNOVATION AND VALUE CREATION

Although it is difficult to get a good, direct measure of the level of innovation and value creation in a country, there are a number of indirect measures and indicators, which suggest that while the United States remains among the world's leaders in activities along the value chain, other countries are advancing rapidly. The following sections review evidence from several perspectives: US performance in the production of manufactured goods and high-tech services, invention, and the country's attractiveness to innovative companies.

How Is the United States Doing in Manufacturing and High-Tech Services?

A useful way to start examining the ability of a country to create value is to look at "value added." This measure captures the amount that a country, company, or other entity contributes to the value of a good or service through the contribution of labor and/or capital inputs. Basically, it represents the price of the good or service produced discounting all (domestic or imported) purchased materials or other external inputs needed to produce it.

Every two years the US National Science Board publishes a report on *Science and Engineering Indicators* that includes value added across industries. The most recent report indicates that the United States' global share of value added in several important industry categories has significantly declined since the early 2000s (NSB 2014). Specifically, the US share of high-tech manufacturing—aircraft, spacecraft, communication products, computers, pharmaceuticals, semiconductors, and technical instruments—dropped from 34 percent in 2002 to 27 percent in 2012. The United States increased the absolute value it contributes in high-tech manufacturing during this period, but not nearly as quickly as China, which increased the value it contributes by more than seven times. It outpaced the United States in computer and office machinery in 2005 and has continued to accelerate in these areas while the corresponding US contributions have stagnated.

Other countries are also starting to surpass the United States in high-tech manufacturing. In 2010 the United States ranked behind Japan in communications equipment and behind the European Union (EU) in technical instruments, and tied with the EU in pharmaceuticals. It remains the world leader in aerospace manufacturing, but there are serious concerns that multiple Asian and Middle East countries will emerge as significant competitors in the near future.

Although US performance in high-tech services has fared better than in high-tech manufacturing, the nation's share of global contributions has declined in this area as well. The United States remains the global leader in business, financial, and communication services, but its share of global value added in these areas fell almost 12 points, from approximately 44 percent to 32 percent in 2001–2012. The value contributed by developing countries, and even developed countries outside of the European Union or Japan (e.g., South Korea, Taiwan, Canada, and Australia), has significantly increased over the past decade, albeit from a lower baseline—from approximately 20 percent in 2001 to 34 percent in 2012 (NSB 2014).

How Is the United States Doing in Invention?

One rough measure of a country's level of value creation is its rate of invention, which can be determined by the number of patents that it applies for or is granted. Again, this is far from an ideal measure—for a number of reasons, including the fact that patents reflect only one aspect of creating value—but it does offer some insights into trends in invention, which is an important part of value creation. US inventors have consistently accounted for roughly half of the patents granted by the US Patent and Trademark Office (USPTO), but since 2003 have gone from just over 50 percent to under 50 percent, with non-US inventors making up the difference (Figure A-1).

A similar picture emerges from a look at "triadic" patents—inventions patented in the United States, the European Union, and Japan. Because it is expensive to apply for patents, inventions patented in all three of these markets are likely to be the most important innovations, economically speaking. From 1999 through 2008 inventors from the each of the three markets accounted for about 30 percent of the total number of triadic patent applications, although their percentages dropped slightly (e.g., from about 32 percent to about 30 percent for the United States) as the percentage of triadic patent applications from the "Asia 8"—India, Indonesia, Malaysia, the Philippines, Singapore, South Korea, Taiwan, and Thailand—rose from about 2 percent to about 4 percent of the total (NSB 2012).

China's contributions are relatively small in these measures because Chinese inventors have made relatively few patent applications in the United States, Europe, or Japan. This does not tell the whole story, however. The number of patents filed by Chinese inventors has risen dramatically over the past decade, but so far primarily in China. In 2012 residents of China accounted for the largest number of patents filed throughout the world and the Chinese Intellectual Property Office accounted for the largest number of applications received by any single IP office (WIPO 2013).

The rapidly growing number of patents filed by Chinese inventors and China's rise in R&D suggest that Chinese patent filings may grow much faster than US patent filings in the coming years and that they may be increasingly for higher-value inventions that rival those from the developed world.

Are the Most Innovative Companies Based in the United States?

Various organizations have attempted to identify the world's most innovative companies using a variety of criteria that yield a broad range of answers. Notwithstanding the diversity of results, it is instructive to examine them to look for trends in the global distribution of these businesses. It is interesting to note that, although the rankings considered all types of companies, those that ranked at the top of each are in manufacturing or high-tech service value chains.

Several rankings simply count the number of patents awarded in a given year to different companies. By that standard, IBM was the world's most innovative company in 2013—for the 21st year in a row (Barinka 2014): it was granted 6,809 US patents, almost 50 percent more than the second-place company, Samsung. Two other US-based companies—Microsoft and Qualcomm—

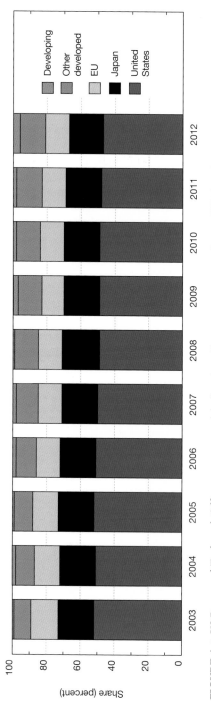

FIGURE A-1 US Patent and Trademark Office patents granted, by location of inventor, 2003–2012. Source: NSB (2014).
Notes: EU = European Union. Technologies are classified by the Patent Board. Patent grants are fractionally allocated among countries on the basis of the proportion of the residences of all named inventors.

were in the top ten, which were dominated by Asian companies. The same was true in 2012, when IBM, Microsoft, and GE were the only US companies in the top ten. The number of US-based companies in the top ten has slowly but steadily declined from four in 2010 and five in 1985.[1]

Innovativeness rankings published by *Forbes* magazine took a completely different tack: they are based on investors' judgments about companies' abilities to create value. This approach draws on work by Hal Gregerson of the international graduate business school INSEAD and Jeff Dyer at Brigham Young University (Gregerson and Dyer 2013). The two researchers calculated what they call an *innovation premium*, a measure of how much investors push up the stock price of a company in anticipation of the company's creating additional value—that is, beyond its current value and what might be projected if the company were not going to innovate further.

The method shows that investors are willing to pay more for companies they expect to be above average in creating value. Based on this criterion *Forbes* ranks the US companies Salesforce.com and Alexion Pharmaceuticals as the two most innovative companies in the world, with four other US companies—Regeneron Pharmaceuticals (5), Amazon.com (6), BioMarin Pharmaceutical (7), and VMware (9)—in the top ten (*Forbes* 2014).

The Boston Consulting Group develops a yearly list of the world's 50 most innovative companies by surveying more than 1,500 senior executives from companies around the world and then combining the executives' ratings with data on revenue growth, three-year shareholder growth, and margin growth. According to the BCG rankings, Apple is the most innovative company in the world, followed by Samsung, Google, Microsoft, Toyota, IBM, Amazon, Ford, BMW, and General Electric, meaning six of the top ten are American companies. A total of 24 US companies made the list of the top 50 (Nisen 2013). This is down from 33 in 2008 (Andrew et al. 2008).

None of these ways of identifying the most innovative companies is ideal. Counting patents tends to tilt the list toward companies in industries that are most patent-heavy, particularly companies in information and communication technology. The method of looking at a company's "innovation premium" relies on investors' perceptions of companies and ends up tilting the list toward smaller, newer companies because it is easier for innovation to make a large difference in their bottom line. The survey method picks out larger companies because those are the ones that more business executives are familiar with. Still, a look at these different lists does suggest a general conclusion: While US companies remain among the most innovative in the world, companies in other countries are catching up.

---

[1] Data available at www.ipo.org (accessed January 23, 2014).

KEY INNOVATION INPUTS IN THE UNITED STATES

The ability of a country to attract and retain businesses along the value chain is influenced by the condition of various inputs that are necessary for innovation. Countries invest in three main categories of these inputs: research and development, education, and infrastructure.

How Is the United States Doing in Research and Development Investments?

In 1999 the United States accounted for 38 percent of all R&D spending around the world. The European Union accounted for 27 percent, and Asia—China, India, Japan, South Korea, Malaysia, Singapore, and Thailand—represented 24 percent. Thus the United States, European Union, and Asia were doing almost 90 percent of the world's research and development (NSB 2012).

Since then the US share of world R&D investment has steadily declined, mainly because China has been ramping up its R&D spending so quickly. According to statistics from the National Science Foundation, by 2009 the US share of the $1.28 trillion in global R&D spending had fallen to 31 percent—a drop of 7 percent in a decade—while the Asian share had risen to 32 percent. The European Union share had also declined, to 23 percent. Much of the growth in the Asian R&D spending came from China, which increased its spending by about 20 percent each year, several times the US rate of growth, albeit from a much lower base. By 2009 China's R&D spending was 12 percent of the world total, and it had surpassed Japan to become the world's second largest investor in research and development (NSB 2012).

The trend has continued. It is estimated that in 2014, $1.62 trillion will be spent on research and development worldwide, of which the United States will account for 31.1 percent, China an estimated 17.5 percent, and all of Asia 39.1 percent. Japan has fallen further behind China, and in 2014 its R&D spending will be 10.2 percent of the world total. Europe will account for 21.7 percent of worldwide R&D spending; Germany, at 5.7 percent, will be the largest European investor in research and development (Figure A-2).

These changes are occurring quickly. In just five years the US share of global R&D dropped from 34 to 31 percent, and the European share from 26 to 22 percent, while the Asian share increased from 33 to nearly 40 percent, and China alone jumped from 10 percent to nearly 18 percent (Grueber and Studt 2013).

These trends are expected to continue through at least 2020. While US R&D spending is expected to increase modestly, Chinese R&D spending is projected to continue its double-digit growth. Assuming the current rates of growth and investment continue, China's total R&D funding is projected to exceed US funding by about 2022 (Grueber and Studt 2013). This is in line

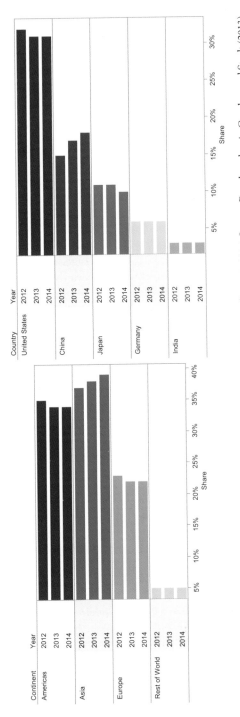

FIGURE A-2 Share of total global research and development spending (actual and estimated), 2012–2014. Source: Based on data in Grueber and Studt (2013).
Note: America, Europe, Asia, and Rest of World include 21, 20, 34, and 36 countries, respectively.

with the country's announced goal of transitioning from an imitation-driven economy to an innovation-driven economy by 2020.

Given the different sizes of countries' economies, perhaps a better way to compare R&D spending is to examine it as a percentage of gross domestic product (GDP). This measure offers a sense of R&D spending *intensity*—how much a country focuses its spending on research and development.

For several decades the United States has consistently spent 2.4–2.8 percent of its GDP on research and development. Over the past decade it generally stayed above 2.6 percent, and in the past few years it remained close to 2.8 percent.

Generally speaking, developed countries spend a much greater percentage of their GDP on research and development than developing countries, but even among developed countries there is large variation. For example, in 2011 South Korea's R&D spending represented 4.0 percent of its total economic output, and Japan's was 3.4 percent. Although the economies of both countries are much smaller than the US economy, they rank high in terms of total R&D expenditures; South Korea's is the fourth largest with $45 billion and Japan's the second largest, after the United States, with $149 billion (OECD 2010).

In 2011 China spent a significantly smaller percentage (about 1.7 percent) of its GDP on research and development (Figure A-3), but that number is growing steadily. The country's R&D spending was only 0.8 percent in 1999 (NSB 2014), it is projected to be 2.0 in 2014, and it seems poised to meet the current five-year plan goal of 2.2 percent by 2015 (Grueber and Studt 2013). As China continues its push to transition to an innovation-driven economy, this number can be expected to increase further, with the result that shortly after 2020 China will devote a greater percentage of its economy to research and development than the United States.

How Is the United States Doing in Science, Technology, Engineering, and Math Education?

An effective system of education in science, technology, engineering, and mathematics (STEM) is crucial for maintaining a country's ability to innovate and create value (see, e.g., NRC 2007). As Atkinson and Mayo (2010, p. 22) wrote in *Refueling the US Innovation Economy*,

> Science- and technology-based innovation is impossible without a workforce educated in science, technology, engineering, and math. As a result, it behooves the United States to support strong science, technology, engineering, and mathematics (STEM) education, especially as our competitors recognize the links between STEM education, greater research, and increased innovation.

By many accounts, the US system of higher education remains the best in

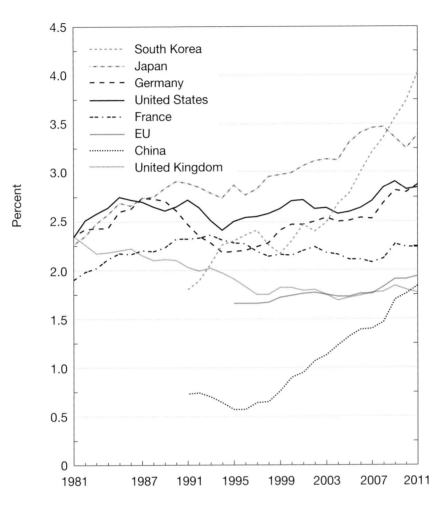
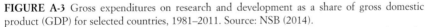

FIGURE A-3 Gross expenditures on research and development as a share of gross domestic product (GDP) for selected countries, 1981–2011. Source: NSB (2014).

Notes: EU = European Union. Data are not available for all countries in all years. Figure shows the top seven R&D-performing countries. Data for the United States reflect international standards for calculating gross expenditures on R&D, which differ slightly from the National Science Foundation's protocol for tallying US total R&D expenditures. Data for Japan since 1996 may not be consistent with earlier data because of changes in methodology.

the world. Universities in the United States typically dominate global rankings of prestigious higher education institutions: in 2011–2012, 18 of the highest-ranking 25 universities and 30 of the top 50 were in the United States (OECD 2011).

Unfortunately, a number of concerns persist about US STEM education among business leaders and negatively affect the perception of the country as an attractive place to locate activities along the value chain. In particular, the quantity of STEM graduates and the quality of K–12 education are cited as leading concerns. These issues are discussed below.

Quantity of Science and Engineering Graduates

The 2012 *Science and Engineering Indicators* show a dramatic increase in the number of students in China earning university degrees in engineering and the natural sciences that are broadly comparable to US baccalaureate degrees. The number of these graduates rose from about 280,000 in 2000 to 1 million in 2008 (Figure A-4). These numbers dwarf the quantity of US graduates in such fields, especially in engineering: In 2008 the number of US natural science degrees awarded was 175,000 and the number of engineering degrees 60,000.

What's more, in the United States, unlike China, a significant percentage of these degrees, especially advanced degrees, are awarded to foreign nationals who may leave the country (Figure A-5). Since 2000, foreign students with temporary visas have earned 39 percent to 48 percent of US doctoral degrees in the natural sciences and engineering. More than half of these students are from China, India, and South Korea (NSB 2012).

Alongside the total number of science and engineering graduates, it is useful to look at their percentage of a country's total population of university graduates. Because China has a much larger population than the United States, it is no surprise that it graduates more students in these fields. However, the United States trails behind not only China but also Japan, South Korea, Germany, and the United Kingdom. The fraction of US university degrees in science or engineering—at 32 percent—pales in comparison to Japan's nearly 60 percent and China's 50 percent.

Ranking of US K–12 Education

Another significant factor affecting the ability of the United States to attract businesses is its K–12 education system. Perceptions of a poor-quality K–12 system are cited as one of America's greatest competitive weaknesses when businesses are considering location decisions (Porter and Rivkin 2012). For example, the OECD Program for International Student Assessment (PISA) ranks the United States 36th out of 65 countries in students' performance in math, 28th in science, and 24th in reading (OECD 2012). Another assessment,

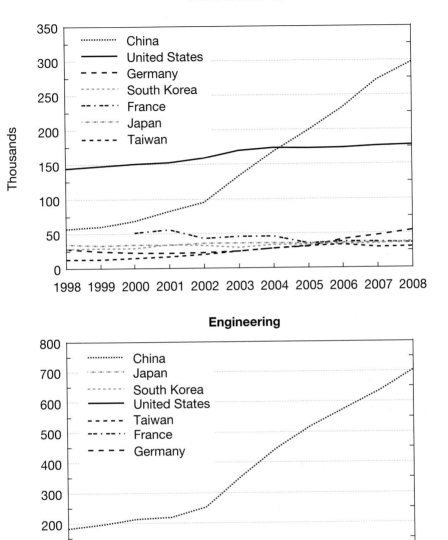

FIGURE A-4 First university degrees in natural sciences and engineering, by selected country, 1998–2008. Source: NSB (2012).
Note: Natural sciences include physical, biological, environmental, agricultural, and computer sciences, and mathematics.

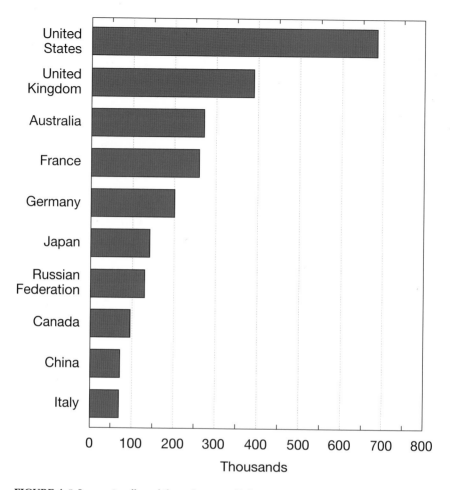

FIGURE A-5 Internationally mobile students enrolled in tertiary education, by selected country, 2010. Source: NSB (2014).
Note: "Mobile students" are defined as those who moved to another country in 2010 with the objective of studying. Data for Canada and the Russian Federation correspond to 2009. Data for Germany exclude advanced research (e.g., doctoral) programs.

the Trends in International Mathematics and Science Study (TIMSS), shows US fourth-graders ranked behind seven other countries, most of them in Asia: Singapore, South Korea, Hong Kong, Taiwan, Japan, Northern Ireland, and Belgium. In science, US fourth-graders lagged behind their counterparts in six other countries: South Korea, Singapore, Finland, Japan, the Russian Federation, and Taiwan (Provasnik et al. 2012).

In 1992 the National Center for Education Statistics released *International*

Mathematics and Science Assessment: What Have We Learned? (Medrich and Griffith 1992), which reported international assessments of students' performance in science and mathematics since the 1960s. The tests painted a uniformly grim picture of the achievement of US students (Medrich and Griffith 1992, p. viii): "The evidence suggests, in general, that students from the United States have fared quite poorly on these assessments, with their scores lagging behind those of students from other developed countries.... Generally the 'best students' in the United States do less well on the international surveys when compared with the 'best students' from other countries."

International STEM Assessments in Context

It is clear that the United States is not doing as well as many other countries, particularly those in Asia, in teaching science and mathematics to students in primary and secondary schools, at least not in terms of producing students who score well on these standardized tests. Many observers have suggested that this comparatively poor performance poses a threat to US competitiveness and prosperity. Indeed, this argument has been made at least since the publication of *A Nation at Risk* (National Commission on Excellence in Education 1983) more than 30 years ago.

A look at the evidence, however, suggests that there is no straightforward connection between performance on these tests and economic competitiveness. The use of international education statistics such as those of PISA does not necessarily represent a fair comparison across countries. It has been noted, for example, that Shanghai has an "economically and culturally elite population with systems in place to make sure that students who may perform poorly are not allowed into public schools" (Loveless 2013, p. x). This skews Shanghai's PISA scores because they do not represent average performance across the population. In contrast, because the United States emphasizes universal primary and secondary education, the US scores may present a somewhat more representative average across the population.[2]

What Is the Condition of US Infrastructure Needed for Value Creation?

In 2014 the World Economic Forum (WEF), in its report on global competitiveness, scored countries around the world on the quality of their transportation, electrification, and telephony (Schwab and Sala-i-Martín 2014). The United States was ranked 19th out of 148 economies. Switzerland, Hong Kong, and Finland were judged to have the best overall infrastructure; the United Arab Emirates was 4th, and Singapore 5th.

---

[2] US scores exclude a not insignificant number of students who drop out of high school before the 9th grade.

Final Observations about US Innovation Inputs

If there is any common trend in these innovation inputs, it is that the United States has serious competition in all of these areas. Several other countries spend more per capita on research and development, for example, and China may soon outspend the United States in terms of total R&D funding. Many other countries score better on multiple measures of science and mathematics education in primary and secondary schools, although comparison of a country as diverse and populous as the United States with homogeneous and small countries is somewhat problematic. And many other countries have higher-quality infrastructure—for telecommunications, Internet, electricity, roads, railroads, and air transportation—than the United States.

All of these factors play a critical role in how well a country can create value, so the United States can no longer take for granted that it will remain the best in the world in this area.

JOBS AND PRODUCTIVITY

As the US Department of Commerce reported in *The Competitiveness and Innovative Capacity of the United States* (DOC 2012, p. 1-4):

> The United States' ability to create jobs has deteriorated during the past decade. Employment increased at an annual rate of just 0.6 percent between the February 2001 and January 2008 employment peaks…. This rate is one-third as fast as the 1.8 annual rate of employment growth between the June 1990 and February 2001 employment peaks. A recent study by McKinsey Global Institute found that the United States has been experiencing increasingly lengthy jobless recoveries [Manyika et al. 2011, p. 1]: "it took roughly 6 months for employment to recover to its prerecession level after each postwar recession through the 1980s, but it took 15 months after the 1990–91 recession and 39 months after the 2001 recession."

Whether or not the decay of employment growth over the past two decades is a direct result of a relative weakening of the United States' ability to create value, the fact remains that the *only* way to create economic growth is to innovate—either by developing a novel product, service, or process that adds value or by putting innovations into widespread practice to improve productivity (Schumpeter 1934). The only way for businesses to stay competitive and provide employment is to continue to create valuable products. Simply put, the economic prosperity of any nation is directly tied to its ability to make value.

Although economic prosperity is often spoken of in the cold language of statistics—employment growth, wage rates, international competitiveness rankings, and so on—it is really about whether people's lives will be better and more comfortable than their parents' lives, as Americans have come to expect.

It is about the sort of society Americans will live in, what sorts of jobs we will have, or whether we will have jobs at all. It is about what sort of future we will create for ourselves and our children.

Fundamental changes in the US economy appear to be looming that underscore the importance for the United States to create new opportunities for innovation, for novel products and services that will generate employment growth.

US ECONOMIC GROWTH AND EMPLOYMENT

GDP growth per employed person in the United States has slowed. This ratio is an important indicator of productivity growth, which is generally considered essential for maintaining or enhancing living standards. Although the recession was certainly a factor, the rate of US productivity growth had been slowing since the early 2000s (Figure A-6); from 1990 to 2000, average growth was 2.0 percent; in 2005–2008 it slowed to 0.9 percent (NSB 2012). Meanwhile, economic growth per employed person in developing economies has exploded, particularly in China and India, which had respective growth rates of 10 percent and 6 percent in 2005–2008 and have continued at these rates through 2012 (NSB 2014).

Until recently employment in the United States was tightly coupled with rising productivity: as workplaces became more efficient and output per worker went up, more jobs were created and wages generally increased. This was the case from at least the 1940s until the 1980s (Brynjolfsson and McAfee 2011). Since then, both wages and the employment-to-population ratio have stagnated despite continued growth in productivity and GDP.

Beginning in the 1980s, median wages stopped growing—job creation kept up with productivity growth but, for the average worker, these jobs did not offer a chance to climb the economic ladder. Then in the early 2000s job creation also stopped growing and it has not turned around since. The United States is creating jobs but not enough to keep up with population growth, and the average American household's income has not improved since 1997. What's more, there's no indication that this picture is likely to change (Brynjolfsson and McAfee 2011).

Thus economic progress and profit growth no longer translate into arguably the most essential achievement that most people strive for: a well-paying job that will allow them to have a higher standard of living.

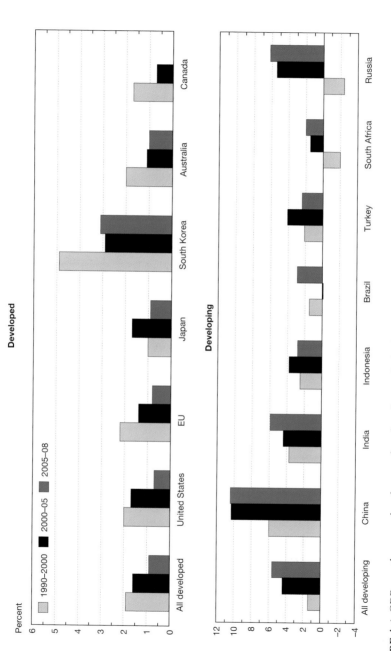

FIGURE A-6 GDP growth per employed person for selected countries/regions, 1990–2008. Source: NSB (2012).
Note: GDP is in 2010 purchasing power parity dollars. The European Union (EU) includes current member countries. China includes Hong Kong. Brazil's growth in 2000–2005 was –0.1 percent.

REFERENCES

Andrew JP, Haanaes K, Michael DC, Sirkin HL, Taylor A. 2008. Innovation 2008: Is the Tide Turning: A BCG Senior Management Survey. Boston: Boston Consulting Group.

Atkinson RD, Mayo M. 2010. Refueling the US Innovation Economy: Fresh Approaches to Science, Technology, Engineering, and Mathematics (STEM) Education. Washington: Information Technology and Innovation Foundation.

Barinka A. 2014. IBM wins most US patents for 21st year in a row. Bloomberg Technology, January 14, www.bloomberg.com/news/2014-01-14/ibm-wins-most-us-patents-for-21st-year-in-a-row.html (accessed January 25, 2014).

Brynjolfsson E, McAfee A. 2011. Race Against the Machine: How the Digital Revolution Is Accelerating Innovation, Driving Productivity and Irreversibly Transforming Employment and the Economy. Digital Frontier Press.

DOC [US Department of Commerce]. 2012. The Competitiveness and Innovative Capacity of the United States. Washington.

Forbes. 2014. The world's most innovative companies. Forbes, August 14. Available at www.forbes.com/innovative-companies/list/ (accessed January 14, 2014).

Gregerson H, Dyer J. 2013. The world's most innovative companies—2013. Available at http://knowledge.insead.edu/innovation/entrepreneurship/the-worlds-most-innovative-companies-2013-2596 (accessed January 14, 2014).

Grueber M, Studt T. 2013. 2014 Global R&D funding forecast. R&D Magazine, December: 1–35.

Loveless T. 2013. The 2013 Brown Center Report on American Education: How Well Are American Students Learning? Washington: Brookings Institution.

Manyika J, Lund S, Auguste B, Mendonca L, Welsh T, Ramiswamy S. 2011. An Economy that Works: Job Creation and America's Future. McKinsey Global Institute. Available at www.mckinsey.com/mgi/publications/us_jobs/pdfs/MGI_us_jobs_full_report.pdf (accessed February 5, 2014).

Medrich EA, Griffith JE. 1992. International Mathematics and Science Assessment: What Have We Learned? Washington: National Center for Education Statistics.

National Commission on Excellence in Education. 1983. A Nation at Risk: The Imperative for Educational Reform. Washington: US Government Printing Office.

Nisen M. 2013. The 50 most innovative companies in the world. Business Insider, September 28. Available at www.businessinsider.com/most-innovative-companies-in-the-world-2013-9 (accessed January 14, 2014).

NRC [National Research Council]. 2007. Rising Above the Gathering Storm: Energizing and Employing America for a Brighter Future. Washington: National Academies Press.

NSB [National Science Board]. 2012. Science and Engineering Indicators 2012. NSB 12-01. Arlington, VA: National Science Foundation.

NSB. 2014. Science and Engineering Indicators 2014. NSB 14-01. Arlington, VA: National Science Foundation.

OECD [Organization for Economic Cooperation and Development]. 2010. Factbook 2010: Economic, Environmental and Social Statistics. Paris. Available at www.oecd-ilibrary.org/economics/oecd-factbook-2010_factbook-2010-en (accessed January 26, 2015).

OECD. 2011. Education at a Glance 2011. "Indicator C3: Who studies abroad and where?" Paris. Available at www.oecd.org/dataoecd/61/2/48631582.pdf (accessed January 23, 2014).

OECD. 2012. PISA Results in Focus: What 15-Year-Olds Know and What They Can Do with What They Know. Paris. Available at www.oecd.org/pisa/keyfindings/pisa-2012-results-overview.pdf (accessed January 14, 2014).

Porter ME, Rivkin JW. 2012. Prosperity at Risk: Findings of Harvard Business School's Survey on US Competitiveness. Cambridge, MA: Harvard Business School.

Provasnik S, Kastberg D, Ferraro D, Lemanski N, Roey S, Jenkins F. 2012. Highlights from TIMSS 2011: Mathematics and Science Achievement of US Fourth- and Eighth-Grade Students in an International Context. NCES 2013-009. Washington: National Center for Education Statistics, Institute of Education Sciences, US Department of Education.

Schumpeter JA. 1934. Capitalism, Socialism, & Democracy. New York: Taylor and Francis. Fifth edition, 1976.

Schwab K, Sala-i-Martín X, eds. 2014. The Global Competitiveness Report, 2013–2014: Full Data Edition. Geneva: World Economic Forum.

WIPO [World Intellectual Property Organization]. 2013. 2013 World Intellectual Property Indicators: Highlights. Geneva. Available at www.wipo.int/export/sites/www/ipstats/en/wipi/2013/pdf/wipo_pub_941_2013_highlights.pdf (accessed January 10, 2014).